MATHEMATICS ACTIVITIES FOR ELEMENTARY SCHOOL TEACHERS
A PROBLEM SOLVING APPROACH
FIFTH EDITION

Dan Dolan
Wesleyan University

Jim Williamson
University of Montana

Mari Muri
Connecticut Department of Education

PEARSON
Addison
Wesley

Boston   San Francisco   New York
London   Toronto   Sydney   Tokyo   Singapore   Madrid
Mexico City   Munich   Paris   Cape Town   Hong Kong   Montreal

Reproduced by Pearson Addison-Wesley from electronic files supplied by the author.

Copyright © 2004 Pearson Education, Inc.
Publishing as Pearson Addison-Wesley, 75 Arlington Street, Boston MA 02116

ISBN   0-321-17413-5

 3 4 5 6 VHG 06 05 04 03

Preface

ABOUT THIS BOOK

Mathematics Activities for Elementary School Teachers provides a hands-on, manipulative based, problem solving approach to learning and teaching elementary mathematics. The activities in the book were developed to correspond to the content in *A Problem Solving Approach to Mathematics for Elementary School Teachers*, Eighth Edition by Rick Billstein, Shlomo Libeskind, Johnny Lott (Pearson Education, Inc. 2004). Although this activities manual was designed to supplement the textbook, it can be used to develop students' understanding of mathematical concepts and skills in a variety of settings:

- mathematics content courses for preservice elementary teachers, grades K-8,
- mathematics methods courses for preservice elementary teachers, grades K-8, and
- inservice courses and professional development workshops for elementary and middle school teachers.

These activities demonstrate an alternative approach to the traditional teaching and learning of mathematics. They are based on a constructivist philosophy and sequenced in a developmentally appropriate manner. The activities can be used to:

- develop mathematical concepts or procedures;
- reinforce a concept that has been previously taught;
- illustrate applications of mathematical concepts in contextual situations; and
- promote individual construction of knowledge of mathematics.

This activities book has been designed primarily for preservice and inservice classes, however, it can also serve as a resource for elementary teachers at various grade levels, K through 8.

A NEW APPROACH

In *Mathematical Knowledge for Primary Teachers*, (Suggate, Goulding, Davis; Fulton Publishers, London, 1998) it states that, "Knowing rules and procedures can be extremely useful and efficient. However, they need to be embedded within a rich relational understanding of the concepts concerned. . . . Someone with mere instrumental understanding of mathematics cannot use and apply it. They are likely only to be able to implement the procedure in contexts which closely resemble the situations in which they were taught the procedure. . . . Their mathematics is rigid and inert. This is hardly the desirable state of mind for a primary teacher who plays a significant part in shaping not only what young children know about mathematics, but also how they feel about it."

"Effective teaching conveys a belief that each student can and is expected to understand mathematics and each will be supported in his or her efforts to accomplish this goal."

Principles and Standards for School Mathematics
NCTM, 2000

The *Principles and Standards for School Mathematics* (National Council of Teachers of Mathematics, 2000) describes knowing mathematics as having the ability to use it in meaningful ways. In the process of learning mathematics, teachers must be involved in doing mathematics–investigating, conjecturing, discussing, and validating–in order to develop confidence in their own mathematical ability and to be able to instill an appreciation of its value in their students. "They (teachers) need to know the ideas with which students often have difficulty and ways to help bridge common misunderstandings."

A Call for Change: Recommendation for the Preparation of Teachers of Mathematics (Mathematical Association of America, 1991) asserts that, "Collegiate mathematics classrooms must become a place where students actively do mathematics rather than simply learn about it." It states further that, "Teaching collegiate mathematics must change to enable learners to grapple with the development of their own mathematical knowledge. . . . the collegiate curriculum in mathematics must be open to new ways of presenting mathematics."

"The kind of experiences teachers provide clearly play a major role in determining the extent and quality of students' learning."

Principles and Standards for School Mathematics
NCTM, 2000

All of these mathematics reform documents call for a basic restructuring of the curriculum, instructional practices, and assessment systems for mathematics in grades K-16, and of the program for the preparation of teachers of mathematics. In her book, *Knowing and Teaching Elementary Mathematics: Teacher's Understanding of Fundamental Mathematics in China and the United States,* (L. Erlbaum Associates Publishers, Mahwah, NJ, 1999) Liping Ma points out that, "the change that we are expecting (from a classroom focusing on procedures to one that includes inquiry) can occur only if we work on changing teachers' knowledge of mathematics."

This transformation necessitates dramatic changes in the ways that prospective teachers learn and later teach mathematics. As noted in the MAA document, *A Call for Change,* alternative methods must be presented to preservice and inservice teachers so that they can learn and practice them while they are in the learning process themselves.

GOALS OF THE BOOK

This book was written to engage preservice and inservice teachers in doing mathematics rather than simply reading about mathematics. It is not intended to be a textbook for a mathematics content course at the collegiate level, or to provide all of the mathematical content necessary for such a course. Rather, it is intended to be used as a companion to a textbook, such as *A Problem Solving Approach to Mathematics for Elementary School Teachers,* to provide hands-on, manipulative-based activities that involve preservice elementary teachers in discovering mathematical concepts, doing real problem solving, and exploring mathematical concepts in interesting, stimulating, and real-world settings.

The content and instructional approach of this activities manual embodies the spirit and intent of the NCTM Standards documents. In the process of engaging students in meaningful mathematical tasks, group work, hands-on activities, and classroom discourse, these materials capture the essence of the first standard for the Professional Development of Teachers in the *Professional Teaching Standards for School Mathematics*, (NCTM, 1991).

Standard 1

Mathematics and mathematics education instructors in preservice and continuing education programs should model good mathematics teaching by

- posing worthwhile mathematical tasks;
- engaging teachers in mathematical discourse;
- enhancing mathematical discourse through the use of a variety of tools, including calculators, computers, and physical and pictorial models;
- creating learning environments that support and encourage mathematical reasoning and teachers' dispositions and abilities to do mathematics;
- expecting and encouraging teachers to take intellectual risks in doing mathematics and to work independently and collaboratively;
- representing mathematics as an ongoing human activity; and
- affirming and supporting full participation and continued study of mathematics by all students.

If elementary teachers learn concepts through a problem-solving activity approach, develop ideas from the concrete level to the abstract level, and connect multiple representations of mathematical ideas, they will gain a broader and deeper understanding of mathematics and begin to construct their own meanings of mathematical concepts. As a result, they will be better prepared to create rich mathematical learning environments in their classroom for students.

The activities pay careful attention to the recommendations regarding mathematics courses for elementary teachers contained in *A Call for Change*. In addition, they promote the major goals of the NCTM Standards documents. These goals state that students should:

- become mathematical problem solvers,
- learn to communicate mathematically,
- learn to reason mathematically,
- become confident in their ability to do mathematics, and
- learn to value mathematics.

"Effective mathematics teaching requires a serious commitment to the development of students' understanding of mathematics. Because students learn by connecting new ideas to prior knowledge, teachers must understand what their students already know. Effective teachers know how to ask questions and plan lessons that reveal students' prior knowledge; they can design experiences and lessons that respond to, and build on, this knowledge."

Principles and Standards for School Mathematics NCTM, 2000

These goals are consciously reflected in every activity in this book. All of the activities have been tested extensively with preservice teachers, teachers in professional development programs, and with elementary students. They have proven to be effective and stimulating with all groups. We hope that as you engage in the activities, you will enjoy exploring mathematics, become confident in your ability to do mathematics, and will come to love it as we do.

Introduction

CONTENT

This book contains activities designed to provide preservice teachers and teachers engaged in professional development programs opportunities to explore mathematical ideas and to develop conceptual understanding using a problem-solving approach and a variety of manipulative materials.

A problem-solving approach to learning enhances the development, reinforcement, and application of mathematical concepts in a way that practicing rote skills cannot. It also engages preservice teachers in learning mathematics in a way we trust they will apply when teaching their students.

For the most part, the activities do not require expensive or single-use materials. Full-color replicas of some commercial materials that can be used at many levels and in a variety of settings are included in the back of the book.

ORGANIZATION

"Students learn mathematics through the experiences that teachers provide. Thus, students' understanding of mathematics, their ability to use it to solve problems, and their confidence in, and disposition toward mathematics are all shaped by the teaching they encounter in school. The improvement of mathematics education for all students requires effective mathematics teaching in all classrooms.

Principles and Standards for School Mathematics NCTM, 2000

The instructional plan that precedes each activity provides direction for the course instructor and includes the following elements:

- **Purpose** outlines the major mathematical concepts that are developed in the activity and describes which of the major objectives—introducing, developing, reinforcing, or applying a concept—the activity is designed to meet.
- **Materials** describes any special equipment or supplies that are needed for the activity. Models of several manipulatives and materials to be duplicated for selected activities are in the appendix.
- **Grouping** describes the classroom setting for the activity. Many activities can be completed on an individual basis. However, we encourage working in pairs or small groups whenever appropriate in order to model students working collaboratively as recommended in *A Call for Change,* the *Principles and Standards for School Mathematics,* and to model how mathematics should be taught in the elementary grades. See the section *Collaborative Learning* following the Introduction.
- **Getting Started** provides an introduction to the activity or explains any preparation necessary before using it. Several activities include more than one section. In some cases, it may not be essential to do all of them. Since the purpose of these activities is not just to teach mathematical concepts, but also to illustrate how the concepts should be taught in the elementary grades, it is important that preservice and inservice teachers keep the following questions in mind and formulate answers to them as they are completing each activity:

A. How could this activity be used with elementary students?
B. At which grade level would the activity be appropriate?
C. Could the activity be used at more than one grade level?
D. What adaptations would be necessary to make the activity suitable for use with elementary students at various grade levels?
E. How would the use of this activity be important in the mathematical development of elementary students?
F. What assessment methods could be used to determine students' understanding?

- **Extensions** present suggestions and ideas for extending the activity to other mathematical topics or making connections between the mathematical concepts in the activity and their application in the real world or other curricular areas. Included are some or all of the following:

  A. additional questions and problems to explore;
  B. questions to extend ideas in selected problems or the entire activity; and
  C. other activities, problems, and information related to the concepts in the activity.

Once the activity has been completed, the instructor should review and extend the activity by facilitating whole class discussion of the results, summarizing the activity, and formalizing the mathematical content.

TIME REQUIREMENTS The activities in this book can replace much of the lecture time that is usually devoted to teaching topics. Many concepts and skills are developed in the activities by working from the concrete level, to the representational, to the abstract level. Through a carefully guided discussion of the results of an activity, students will develop their own mental constructs of the concepts being presented.

APPENDICES
- Manipulatives. Pages A-1 through A-15 are printed one-sided on heavy stock and in color so that you can cut them out for use with many activities. We suggest that you store each individual set in plastic zip-lock bags or small plastic storage containers.
- Activity Masters. Pages A-17 through A-51 consist of models, activity recording sheets, graph paper, nets for polyhedrons, activity cards, etc. that are to be used as part of selected activities. They should be photocopied for use with an activity and the master retained as it may be used in a later activity.

Collaborative Learning

WHAT IS IT?　Collaborative learning, also described as cooperative learning or cooperative grouping, is any setting where individuals come together to share their talents to solve a problem. Teachers select groups of students to work together based on a variety of factors.

A teacher may select students to work together in order to differentiate instruction. Students of like ability work together to solve problems she has selected based on each group's ability level. Each group works on a problem that will challenge its ability.

However, this grouping strategy does not take into account the various learning styles or mathematical strengths of students. Arranging students within a group according to differentiated mathematical abilities or different learning styles are also effective methods of getting students to work collaboratively. All grouping strategies require that the teacher must know her students well in order to make the best selections.

These latter grouping strategies allow students to share their diverse ways of thinking with other group members. This is beneficial to students of all ability levels. Brighter students become aware of alternative ways of looking at and solving problems. The lesser ability students are exposed to more sophisticated problem solving methods of their peers, and average students learn from both kinds of students. All students, regardless of ability level or learning style, have the opportunity to assimilate a variety of strategies into their problem solving repertoire.

WHY USE IT?　Collaborative learning can improve students' academic achievement, help them build a greater sense of self-confidence, and enhance their social skills. Perhaps the most compelling reason for engaging students in collaborative problem solving is that it will prepare them for the world of work. Successful corporations rely on a team approach. A manufacturing team developing a new aircraft engine works collaboratively. Each team member has an assigned role based on his or her background and specific skills.

The team meets frequently to confer and discuss problems they face. Group discussions promote diverse thinking and problem solving strategies. In the end, the product produced belongs to all who contributed to its development, so each team member shares in the pride of the completed product. Each person learns from other team members, an experience that would not happen if each person worked in isolation.

WHAT DOES IT LOOK LIKE?

Researchers have analyzed and described a variety of grouping strategies to promote collaborative learning. No one method will work in all cases, yet the interdependence of group members' work is a critical factor in the success of collaborative learning. Teachers need to select group members based on their specific skills and strengths and to assure that the group has enough talent among its members to accomplish the task. The group's responsibility is to take advantage of what each person contributes to the solution of the problem.

Group size can vary. Students can work effectively in pairs. The pair may be of different ability levels (one brighter and one slower student) or have different mathematical strengths (one student with strong spatial sense and the other with good numerical skills). Students may have different learning styles (one kinesthetic learner and the other auditory). These students can pool their talents to solve a problem.

Paired work is the basis for a cooperative strategy known as think, pair, share. The teacher poses a problem; students are asked to think about the problem and then share their thinking with a classmate. One student from each pair then reports to the class. For an excellent example of this method, see the Japanese eighth grade mathematics lesson on geometry contained in *The TIMSS Videotape Classroom Study: Methods and Findings from an Exploratory Research Project on Eighth Grade Mathematics Instruction in Germany, Japan, and the United States*.

Group size can also range from three to five students. Students are chosen for a group based on the various strengths or abilities described above. When group size is larger than two, specific roles are assigned to ensure that each student has an opportunity to participate. Roles may include recorder, questioner, materials manager, reporter, etc. Upon completion of the task, each group member is responsible for knowing the solution to the problem.

CHALLENGES FOR THE TEACHER

To facilitate the collaborative learning process, the teacher needs to move around the class and ask strategic questions to keep groups focused and moving ahead without prompting too directly. The teacher also needs to be skillful in encouraging reluctant students to become part of the group effort. All members are encouraged to ask questions of each other. Often, one student is able to answer a question that another has, thus eliminating the need to ask the teacher. When no one in the group is able to answer a particular question, a group designee confers with the teacher and brings the answer back to the group.

The teacher is to maintain individual accountability while working in groups. A commonly used method is to assign a number (one–four) to each member. Once the problem has been solved, the teacher calls on student number three in each group to explain their solution. On another problem, the teacher may call on all the number ones. This reporting method means that each member of the group must be prepared to explain the group's solution strategy.

Lastly, the teacher needs to make decisions about student accountability. Will students work on the test collaboratively or individually? Will she assign a grade for the group, individual grades, or a combination of both?

We urge the users of this activity book to explore the various grouping strategies suggested for each lesson. Keep in mind the various ways of grouping students and ways of promoting full involvement of all participants in the group. We believe working collaboratively will promote greater understanding of the mathematical concepts presented in this book and enhance students' self-confidence in solving problems.

For further reading on Collaborative Learning:

Blueprints for a Collaborative Classroom, Developmental Studies Center, 1997.

Cohen, Elizabeth G., *Designing Groupwork*, Teachers College Press, 1994.

Shulman, Judith, et. al, *Groupwork in Diverse Classrooms*, Teachers College Press, 1998.

Stigler, James W, et al, *The TIMSS Videotape Classroom Study: Methods and Findings from an Exploratory Research Project on Eighth Grade Mathematics Instruction in Germany, Japan, and the Untied States*, U.S. Department of Education, Washington, D.C. 1999

Tomlinson, Carol Ann, *The Differentiated Classroom*, Association for Supervision and Curriculum Development, 1999.

About the Authors

DAN DOLAN Dan Dolan is the Director of the Project to Improve Mastery in Mathematics and Science (PIMMS) at Wesleyan University in Connecticut. PIMMS is the premier professional development program for teachers of mathematics and science in CT. From 1981-1991, he was the State Mathematics and Computer Education Specialist in the Montana State Office of Public Instruction. He has been actively involved in education for 40 years and has taught mathematics at all levels from grade six through university teacher education. He has authored numerous articles and co-authored three books. He presents workshops and conference sessions throughout the country and serves as an editorial consultant for several publishers. He served on the Mathematical Sciences Education Board, the Board of Directors of the National Council of Teachers of Mathematics, and was a member of the 5-8 writing group for the NCTM *Curriculum and Evaluation Standards for School Mathematics*. He has served on various NCTM committees and chaired the editorial panel for the *Student Math Notes*.

MARI MURI Mari Muri is a Mathematics Consultant for the Connecticut State Department of Education and a co-principal investigator of the CT State Systemic Initiative. She plays a principal role in the design and implementation of the CT Mathematics Assessment Program at grades 4, 6, 8, and 10. The performance assessment components of these tests are a model of the type of assessment called for in the NCTM *Assessment Standards* and other mathematics reform documents. Prior to her current position, she was the Mathematics Instructional Consultant for the Killingly (CT) Public Schools. She has been involved in education for 21 years and has taught mathematics at the elementary and university teacher education levels. She served on the Mathematical Sciences Education Board, was a member of the writing team for the NCTM *Assessment Standards,* served as president of the Association of State Supervisors of Mathematics, and as a member of the editorial panel (one year as chair) for NCTM's *Mathematics Teaching in the Middle School.*

JIM WILLIAMSON Jim Williamson is a Visiting Assistant Professor of Mathematics at the University of Montana where he teaches courses for prospective teachers and works on the Show-Me Project, a National Science Foundation (NSF) funded project supporting dissemination and implementation of standards-based middle grades mathematics curricula. Jim also chaired the writing team for the *Six Through Eight Mathematics (STEM) Project*, which was funded by NSF and developed an integrated mathematics curriculum for grades 6-8.

Before joining the STEM Project, Jim served for one year as the Interim Mathematics and Computer Education Specialist in the Montana Office of Public Instruction, as a Visiting Instructor of Mathematics at Montana State University for one year, and as the Mathematics Specialist for the Billings (MT) public schools for four years. He has been involved in mathematics education for 30 years and has taught mathematics at all levels from fourth grade through university. He received a Presidential Award for Excellence in Mathematics Teaching in 1984. While teaching in Columbus, Montana, he co-directed *Math Lab Curriculum for Junior High School*. He has authored several articles, co-authored six books, and presented staff development workshops and conference sessions at meetings throughout the country. He served on the NCTM committee that developed tests and coaching materials for the MATHCOUNTS competitions and chaired the committee for the 1989 competition.

Contents

Chapter 1
An Introduction to Problem Solving

Solving a problem is *to search for some means* to attain some clearly conceived but not immediately attainable solution. If the solution by its simple presence does not instantaneously suggest the means, we have to search for the means, reflecting consciously how to find the solution.

To solve a problem is to find such a means.

George Polya

"A teacher of mathematics has a great opportunity. If the teacher fills the allotted time with drilling students with routine operations, the teacher kills their interest, hampers their intellectual development, and misuses the opportunity. But if the teacher challenges the curiosity of the students by setting them problems proportionate to their knowledge, and helps them solve their problems with stimulating questions, the teacher may give them a taste for, and some means of independent thinking."

George Polya
How to Solve It, 1957

"Problem solving means engaging in a task for which the solution method is not known in advance. In order to find a solution, students must draw on their knowledge, and through this process, they will often develop new mathematical understandings. Solving problems is not only a goal of learning mathematics but also a major means of doing so. Students should have frequent opportunities to formulate, grapple with, and solve complex problems that require a significant amount of effort and should then be encouraged to reflect on their thinking."

—*Principles and Standards for School Mathematics*

The activities in this chapter are designed to help you improve your problem-solving skills. Good problem solvers need to know a variety of techniques for solving problems, so the primary focus of the activities is on developing and applying a variety of problem-solving strategies: *look for a pattern, make a table, use logical reasoning, make a model, and simplify the problem.*

As you complete the activities, you will learn to make conjectures based on observations and data, to verify and generalize the conjectures, and to communicate your results to others. This is the essence of the problem-solving process. The problem-solving strategies and processes are the tools you will use throughout this book to explore and develop mathematical concepts.

Activity 1: Pictorial Patterns

PURPOSE   Identify and extend pictorial patterns.

GROUPING   Work individually or in groups of 2 or 3.

GETTING STARTED   Look for a pattern in the successive pictures. Then draw pictures in the blanks to extend the sequence. Briefly explain the pattern you used.

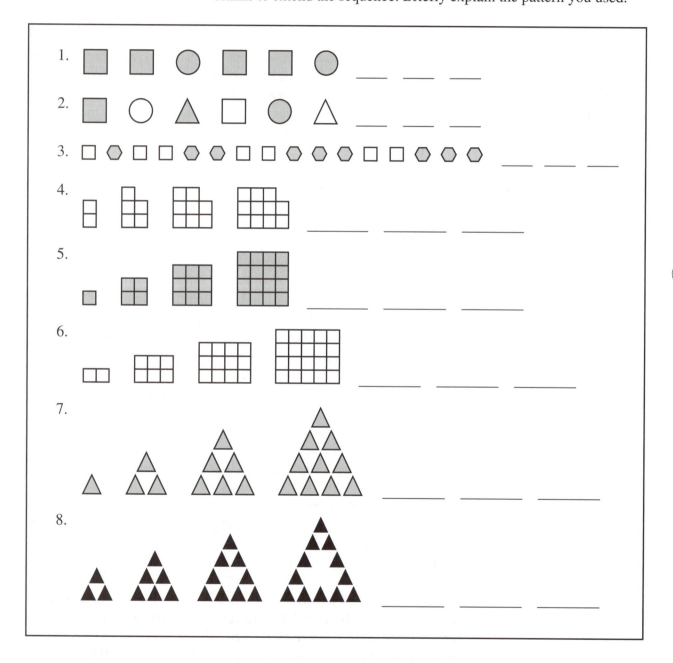

EXTENSIONS   Make up several pictorial patterns of your own and give them to a classmate to extend.

Activity 2: Number Patterns

PURPOSE   Identify and extend numerical patterns and explore relationships between the terms of sequences and the term numbers.

GROUPING   Work individually or in groups of 2 or 3.

GETTING STARTED   Fill in the blanks with the numbers that complete the sequence, and briefly explain the rule you used. In some cases, an intermediate term or the last term is given so that you can check your work.

1.   4, 0, 6, 4, 0, 6, 4, 0, ___, ___, ___, ___, ___, ___

2.   3, 4, 3, 4, 5, 3, 4, 5, ___, ___, ___, ___, ___, 7

3.   1, 2, 1, 1, 2, 3, 2, 1, 1, 2, 3, 4, 3, ___, ___, 1, 2, ___, ___, ___, ___, ___, 2

4.   2, 5, 8, 11, ___, ___, ___, ___, ___, 29

5.   8, 13, 18, 23, ___, ___, ___, ___, ___, ___

6.   53, 46, 39, 32, ___, ___, ___, 4, ___, ___

7.   2, 4, 7, 11, ___, ___, ___, ___, ___, 56

8.   4, 7, 12, 19, ___, ___, ___, ___, ___, ___

9.   2, 3, 5, 5, 8, 7, 11, ___, ___, ___, 17, ___, ___, ___

10.   2, 4, 8, 16, ___, ___, ___, 256, ___, ___

11.   729, 243, 81, 27, ___, ___, ___, ___, ___, $\frac{1}{27}$

12.   2, 3, 5, 8, 13, 21, ___, ___, ___, 144, ___, ___

Use the given rule to determine the first eight terms of each sequence.

Example: Each term is the term number times 4, plus 1.

First Term	Second Term	Third Term	Fourth Term
$1 \times 4 + 1$	$2 \times 4 + 1$	$3 \times 4 + 1$	$4 \times 4 + 1$
5	9	13	17

1. Each term is 3 times the term number, plus 2.

1	2	3	4	5	6	7	8	9
___	8	___	___	___	___	___	___	29

2. Each term is 2 less than 6 times the term number.

1	2	3	4	5	6	7	8	9
___	10	___	___	___	___	___	___	52

3. Each term is the term number times ⁻3, plus 47.

1	2	3	4	5	6	7	8	9
___	___	38	___	___	___	___	___	20

4. Each term is 2 times the square of the term number, plus 5.

1	2	3	4	5	6	7	8	9
___	___	___	___	___	___	___	___	167

5. Each term is the term number times the next term number.

1	2	3	4	5	6	7	8	9
___	___	___	___	___	___	___	___	90

6. Find the difference between successive terms in Exercises 1–3. What do you notice? Does the same thing occur in Exercises 4 and 5? Explain.

7. If you did not know the rules for the sequences in Exercises 4 and 5, how could you find the next five terms of each sequence?

EXTENSIONS   Make up your own rules for generating sequences. Be sure to include increasing and decreasing sequences in which the difference between successive terms is constant as well as some rules in which the differences are not constant. Give them to a classmate to solve.

Activity 3: What's the Rule?

PURPOSE   Develop a procedure for determining a rule that describes the general term of an arithmetic sequence.

GROUPING   Work individually or in pairs.

GETTING STARTED   Fill in each blank to discover a method for determining the rule that generates an arithmetic sequence.

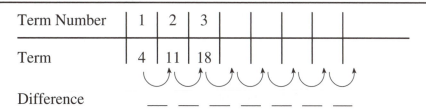

Term Number	1	2	3					
Term	4	11	18					
Difference		__	__	__	__	__	__	__

What is the constant difference? _____

Term Number		Constant Difference			What Was Done?		To Get	
1	×	7	→	7	_____	=	4	First Term
2	×	7	→	____	_____	=	11	Second Term
3	×	____	→	____	_____	=	____	Third Term
10	×	____	→	____	_____	=	____	Tenth Term
50	×	____	→	____	_____	=	____	Fiftieth Term

Write a sentence, like those in Activity 2, that states a rule for generating the terms in the sequence.

Write the rule as an equation.

In each of the following sequences:
- Fill in the missing numbers.
- Find a rule that generates the terms in the sequence.
- Determine the 25th and 100th terms of the sequence.

	Sequence	Rule	25th Term	100th Term
1.	9, 13, 17, 21, ____, ____, ____, ____, …	_____	____	____
2.	2, 9, 16, 23, ____, ____, ____, ____, …	_____	____	____
3.	–3, –1, 1, 3, ____, ____, ____, ____, …	_____	____	____
4.	98, 96, 94, 92, ____, ____, ____, ____, …	_____	____	____
5.	77, 74, 71, 68, ____, ____, ____, ____, …	_____	____	____

Activity 4: Polygons and Diagonals

PURPOSE	Use differences to determine sequences in which the difference between successive terms is not constant.
GROUPING	Work individually or in pairs.
GETTING STARTED	Differences can often be used to analyze and extend sequences even if the difference between terms is not constant. Work the following exercises to learn how.

How many diagonals can be drawn in a dodecagon, a 12-sided polygon?

1. Draw the diagonals in each of the following polygons and record the results in the table. The first two have been done for you.

 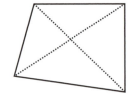 

Number of Sides	3	4								
Number of Diagonals	0	2								

Difference between successive numbers of diagonals   _2_ ___ ___ ___ ___ ___ ___ ___ ___

2. Find the difference between successive terms in the sequence for the number of diagonals. Are the differences constant?

3. Look for a pattern of change in the differences. Continue that pattern to get the next terms in the sequence for the number of diagonals.

The problem of finding the number of diagonals in a dodecagon shows that in order to extend a sequence it is sometimes necessary to look for a pattern in the differences between successive terms. Complete the following exercises to see how this can be done.

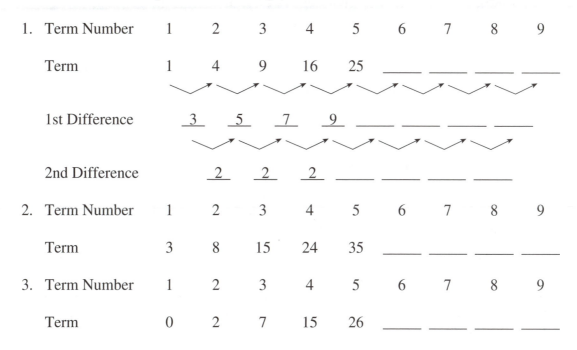

1. Term Number   1   2   3   4   5   6   7   8   9

   Term   1   4   9   16   25   ___   ___   ___   ___

   1st Difference   3   5   7   9   ___   ___   ___   ___

   2nd Difference   2   2   2   ___   ___   ___   ___

2. Term Number   1   2   3   4   5   6   7   8   9

   Term   3   8   15   24   35   ___   ___   ___   ___

3. Term Number   1   2   3   4   5   6   7   8   9

   Term   0   2   7   15   26   ___   ___   ___   ___

---

1. Term Number   1   2   3   4   5   6   7   8   9

   Term   2   10   30   68   130   ___   ___   ___   ___

   1st Difference   8   ___   ___   62   ___   ___   ___   ___

   2nd Difference   12   ___   24   ___   ___   ___   ___

   3rd Difference   6   ___   ___   ___   ___   ___

2. Term Number   1   2   3   4   5   6   7   8   9

   Term   2   4   8   16   ___   ___   ___   ___   186

3. Compare the sequence in the previous exercise to the sequence in Exercise 10 of Activity 2. What can you conclude from these two exercises?

Activity 5: Paper Powers

PURPOSE   Develop an understanding of exponents and exponential change.

MATERIALS   Sheets of newsprint, rulers, and calculators (A scientific model is best for this activity.)

GROUPING   Work individually or in pairs. This may be done as a class activity directed by the instructor.

1. Estimate the number of times you think you can fold a sheet of newsprint if you continue to fold the result in half each time. _____

2. Now, fold the sheet in half as many times as you can. After each fold, count the number of layers of paper and record the result for the **Number of Layers** in **Standard Form** in the table. What happens to the number of layers after each fold?

3. Now record the **Number of Layers** in **Factored Form** and in **Exponential Form** in the table. Begin recording after one fold.

Folds	0	1	2	3	4	5	6	7	8
No. Layers (Std. Form)	1	2							
No. Layers (Fact. Form)			2×2						
No. Layers (Exp. Form)				2^3					
Approx. Height (cm)								1.0	

4. How did your estimate for the number of folds in Exercise 1 compare to the actual number of folds you were able to make?

5. Examine the pattern of entries as you go from right to left in the Exp. Form row. If the pattern continues, what will be the correct entries for 1 fold? _____ 0 folds? _____

6. If a large sheet of newsprint is folded seven times as described above, the thickness is approximately 1.0 cm. Use this number to determine other entries for **Height of the Stack** in the table.

7. Use your calculator to extend the table and determine the number of layers needed to approximate your height.

   Your height (cm) _____ No. of layers _____ Height (from table) _____

8. a. If a sheet could be folded 30 times, would the stack reach the top of the Sear's Tower, _____ an orbiting satellite, _____ the moon? _____
               Y/N   Y/N   Y/N

   b. Use the pattern in the table to determine the height after 30 folds. Describe how your answer compares to the height of the Sear's Tower, the distance to an orbiting satellite, and the distance from Earth to the moon.

Activity 6: The King's Problem

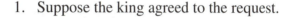

PURPOSE Apply problem-solving strategies to develop an understanding of exponential growth.

MATERIALS Rice and measuring tools

GROUPING Work individually or in groups of 2 or 3.

GETTING STARTED Legend has it that when the inventor of the game of chess explained the game to his king, the king was so delighted he asked the man what gift he would like as a reward.

"My wants are simple," the man replied. "If you but give me one grain of rice for the first square on the playing board, two for the second, four for the third, and so on for all sixty-four squares, doubling the number of grains each time, I will be satisfied."

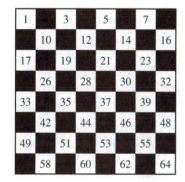

1. Suppose the king agreed to the request.

    a. How many grains of rice would be the inventor receive? **Hint:** How would the number of grains of rice on the seventh square compare to the total number of grains on the first six squares?

    b. How would the total number of grains of rice on the black squares compare to the total number of grains on the white squares?

2. a. How much would the number of grains of rice you found in Exercise 1(a) weigh?

    b. How many bushels of rice would this be?

    c. How large would a building need to be to hold the rice? (Make a sketch of the building and label its dimensions.)

    d. At today's prices, what would the retail value of the rice be?

3. Consult an almanac or the internet to answer the following.

    a. Does the United States produce enough rice in one year to satisfy the inventor's request? Explain.

    b. Is enough rice produced in the world in one year to satisfy the inventor's request? Explain.

    c. How long would it take to produce the needed rice?

Activity 7: When You Don't Know What to Do

PURPOSE	Introduce a four-step approach to solving problems.
MATERIALS	Colored squares (page A-7) and half-centimeter graph paper (page A-48)
GROUPING	Work individually or in pairs.
GETTING STARTED	Problem solving has been described as "what you do when you don't know what to do." As you investigate the following problem, you will learn a four-step approach that may help you solve problems.

Eighteen squares can be arranged into two congruent staircases in which no squares overlap and each step up contains exactly one less square than the step below it. One way this can be done is shown at the left.

Eight squares cannot be arranged into two staircases in this way.

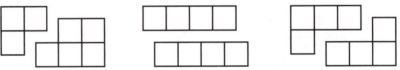

Non-congruent   Not staircases   Step up has two less
staircases   squares than the one below

What other numbers of squares can be arranged into two congruent staircases in this way?

UNDERSTAND THE PROBLEM

The first step in the four-step approach is to make sure you understand the problem. Among other things, this involves reading the problem carefully, sometimes several times, to be certain you understand what the question is. You must also identify the information needed to solve the problem and determine whether any of it is missing.

1. Describe what is meant by a staircase in this problem.

2. What does it mean for shapes to be congruent?

3. Arrange 30 squares into two congruent staircases. How many different ways can you do it?

4. Restate the problem in your own words.

5. How did answering these questions help you understand the problem?

MAKE A PLAN

Once you are sure you understand the problem, the next step is to *make a plan* for solving it. This often involves choosing a problem-solving strategy or strategies to use in solving the problem.

1. Describe how the problem could be modeled using graph paper or squares.

2. What strategies other than *make a model* might be useful in solving the problem.

3. Describe how you would attempt to solve the problem.

CARRY OUT THE PLAN

Carrying out the plan involves using your chosen strategies to attempt to solve the problem. If a particular approach doesn't work, you may need to alter your plan.

1. Carry out your plan for solving the problem.

LOOK BACK

The final step is to *look back* over your solution not just to check your work, but to see what you have learned from solving the problem. Looking back involves checking that your answer is reasonable and that it answers the question that was asked. It also involves looking for other ways to solve the problem and looking for connections to other problems or mathematical ideas.

1. Does your solution include a way to test whether or not a given number of squares can be arranged into two congruent staircases? If not, try to find a way.

2. Try to verify your solution by solving the problem a different way.

3. a. How many different ways can 120 squares be arranged into two congruent staircases?

   b. How can you determine how many different ways a given number of squares can be arranged into congruent staircases?

4. How is the staircase problem related to finding the whole number factors of a number? the odd and even factors?

Activity 8: An Ancient Game

PURPOSE Introduce the work backward problem-solving strategy.

MATERIALS One calculator for each pair of students

GROUPING Work in pairs.

GETTING STARTED This is a version of a game called NIM. It is a game for two players. Beginning with 17, the players alternate turns subtracting 1, 2, or 3 from the number on the calculator display. The first player to make the display read 0 is the winner.

SAMPLE GAME

Player	Keys Pressed	Display
A	17 − 3 =	14
B	− 2 =	12
A	− 3 =	9
B	− 1 =	8
A	− 3 =	5
B	− 3 =	2
A	− 2 =	0

Player A wins!

1. Play the game several times.

2. a. In the Sample Game, what number could Player B have subtracted on his/her last turn and been sure to win the game? Explain.

   b. Find a strategy for winning the game.

3. There are many variations of this game.

   a. Try starting with 25 and subtracting 1, 2, 3, or 4. How does this change the strategy for winning the game?

   b. What is the strategy for winning the following version of NIM? Start with 47 and alternate turns subtracting 3, 5, or 7 from the number on the display. The first player to get a number less than or equal to 0 on the display is the winner.

Activity 9: Ten People in a Canoe

PURPOSE   Introduce the simplify problem-solving strategy and apply the make a table, make a model, and patterns strategies.

MATERIALS   Five each of two different-colored squares (page A-7) for each group

GROUPING   Work individually or in groups of 2 or 3.

Ten people are fishing from a canoe. The seats in the canoe are just wide enough for one person to sit on, and the center seat is empty. The five people in the front of the canoe want to change seats and fish from the back of the canoe, and the five people in the back of the canoe want to fish from the front. Because the canoe is so narrow, only one person may move at a time. A person changing seats may move to the next empty seat, or step over one other person to reach an empty seat. Any other move will capsize the canoe.

What is the minimum number of moves needed to exchange the five people in the front with the five in the back?

HINT: Sometimes, the best approach to solving a problem is to simplify it by considering easier cases of the same problem. Use squares of two different colors to represent the people in the canoe and a model like the one below to represent the seats.

Simplify the problem by solving easier cases. The solution to the problem for two people in the canoe is shown below.

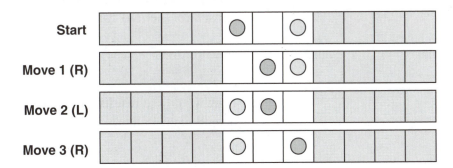

1. Solve the problem for four people. Record the results in the table below.

Number of People	2	4	6	8	10
Number of Pairs	1	2	3	4	5
Minimum Number of Moves	3				
Sequence of Moves	RLR				

2. Complete the table. Look for two patterns, one for how the moves should be made and one for the minimum number of moves.

3. Describe the patterns you found.

4. How many people in the canoe would produce the following sequence of moves?

   R LL RRR LLLL RRRRR LLLLLL RRRRRR LLLLLL RRRRR LLLL RRR LL R

5. If 30 people were in the canoe, how many moves would be needed for them to change places?

6. How would the results change if there was an odd number of people in the canoe?

EXTENSIONS   **The Legend of the Tower of Brahma**

It is said that in a temple at Benares, India, the priests work continuously moving golden disks from one diamond needle to another. It seems that when the world was created, the priests of Benares were given three diamond needles and 64 golden disks. The priests were told that they were to place the disks on one of the needles in decreasing order of size and then move the whole pile to one of the other two needles, moving only one disk at a time and never placing a larger disk on top of a smaller one. According to the legend, God told the priests, "When you finish moving the pile, the world will end."

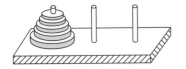

We can simulate the priests' problem by using coins, Cuisenaire rods, or different-sized squares cut from paper to represent the disks. Each peg can be represented by a square in a model like the one below.

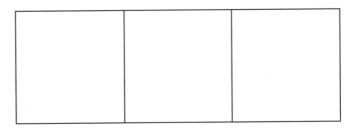

Stack the objects in one of the squares in decreasing order of size. The goal is to move the stack of objects from one square to another in the fewest possible moves. There are two rules: (a) only one object may be moved at a time, and (b) a larger object may never be placed on top of a smaller one.

1. What is the minimum number of moves required to move five objects from one square to another? **HINT:** Look for two patterns, as in the previous problem.

2. Suppose the priests move one disk every second without stopping. How long will it take them to move:

    a. 10 disks?   b. 30 disks?

    c. 50 disks?   d. All 64 disks?

Activity 10: Missing Person Reports

PURPOSE   Use the process of elimination to identify which person is being described by a set of clues.

MATERIALS   One set of attribute people pieces (page A-1) per person

GROUPING   One clue reader, the "sergeant"; groups of 3–8 students, the "detectives"

GETTING STARTED   Each participant lays out a set of attribute people pieces. The sergeant reads the clues on a Missing Person Report and the detectives try to identify the person described by the clues. Each detective holds up the picture of the person he or she identified for verification by the sergeant.

Missing Person Report 1:

This person is large.

He is wearing a dotted jacket.

The jacket is red and makes this clown look very funny.

Missing Person Report 2:

This person is a hobo.

He was last seen wearing a blue jacket.

The jacket has stripes.

This person is large.

Missing Person Report 3:

This person was last seen wearing red.

He is small.

He does not have stripes on his jacket.

He is a hobo.

Missing Person Report 4:

This person is not a hobo.

This person is not small.

This person is not wearing stripes.

This person is not wearing blue or red.

Missing Person Report 5:

This small person is not a clown.

He is wearing stripes of his favorite color.

He is not fond of green or red.

Missing Person Report 6:

This person is not a hobo.

This person is large.

The dots on his jacket are small and red.

EXTENSIONS   Write additional missing person reports similar to the ones above and have a classmate solve them.

Activity 11: Who Dunit?

PURPOSE   Introduce the elimination problem-solving strategy.

GROUPING   Work individually or in groups of 3 or 4.

Inspector Bob E. Sleuth of the London Police Force is investigating a £1 million robbery at the Two Dot Diamond Exchange. The following suspects are in custody:

- "Green-faced" Larry. He gets so car sick, the police had to walk him to the station.
- "Gun Shy" Gordon. He has been afraid of guns since he shot off his big toe as a boy.
- "Loud Mouth" Louise. She is so shy, she leaves her Aunt Jane's house only at night to rent "Wild World of Wrestling" videos at the corner store.
- "Tombstone" Teri. She works the graveyard shift running a forklift at a warehouse.
- "Lefty" McCoy. He lost his left arm in a demolition derby accident.

Use these clues to help Inspector Sleuth solve the crime.

a.   The salesclerk told police the robber had a large handgun.
b.   A waiting taxi whisked the robber away.
c.   The robber wore a large trench coat and a ski mask.
d.   The robber clowned around in front of the security cameras.
e.   The manager said the robber nervously twiddled his or her thumbs while the clerk stuffed diamonds into some sacks.

Who should be booked and held over for trial?

EXTENSIONS   Explain how the clues helped you to eliminate the innocent suspects and to determine the likely suspect.

"How often have I said to you that when you have eliminated the impossible, whatever remains, however improbable, must be the truth?"

Sherlock Holmes, *The Sign of Four*

Activity 12: What's the Number?

PURPOSE   Apply the elimination problem-solving strategy.

GROUPING   Work individually or in groups of 3 or 4.

GETTING STARTED   Use the process of elimination to solve the following number puzzles. Keep a record of the order in which you used the clues and the reasons you used them in that order.

Circle the number below that is described by the following clues.

a. The sum of the digits is 14.
b. The number is a multiple of 5.
c. The number is in the thousands.
d. The number is not odd.
e. The number is less than 2411.

2660   2570   905
1580   1058   1922
1355   1455   770
2290   2435   1770
1832   860   1680

Solve the following number riddle.

a. I am a positive integer.
b. All my digits are odd.
c. I am equal to the sum of the cubes of my digits.
d. I am less than 300.

Who am I? _____

Rebecca has a collection of basketball cards. When she puts them in piles of two, she has one card left over. When she puts them in piles of three or four, there is also one card left over, but when she puts them in piles of five there are no cards left over.

If Rebecca has fewer than 100 basketball cards, what are the possible numbers of cards she could have?

Activity 13: Eliminate the Impossible

PURPOSE   Introduce the method of indirect reasoning.

GROUPING   Work individually or in groups of 2 or 3.

Andrea was visiting her Uncle Ralph, who has a large gumball collection. When she asked if she could have some, he said yes, if she could solve a problem for him. He told her that he has three jars, each covered so that no one could see the color of the gumballs. One jar is labeled red, the second green, and the third red-green. However, he said, no jar has the correct label on it. She could reach into one jar and take one gumball. Then she had to tell him the correct color of the gumballs in each jar. She reached into the jar labeled red-green and pulled out a red gumball.

1. Are there any green gumballs in that jar? Why?

2. What is the correct label for the jar labeled red-green? Explain your answer.

3. Can the jar labeled red contain red and green gumballs? Why?

4. What are the correct labels for each of the jars?

Jorge claims that he has a certain combination of U.S. coins and he cannot make change for a dollar, half dollar, quarter, dime, or nickel. Is this possible? If so, what is the greatest amount of money Jorge could have, and what coins would they be? He does not have any dollar coins.

Total amount: _____

Coins: _____

Students in the fifth grade were playing a trivia game involving states, state birds, and state flowers. They knew that in Alaska, Alabama, Oklahoma, and Minnesota, the flowers are the camellia, forget-me-not, pink-and-white lady's slipper, and mistletoe. The state birds are the common loon, yellowhammer, willow ptarmigan, and scissor-tailed flycatcher. No one knew which bird or flower matched which state. They called the library and received the following clues. Use the clues to complete the table below.

a. The flycatcher loves to nest in the mistletoe.
b. The forget-me-not is from the northernmost state.
c. Loons and lady's slippers go together, but Minnesota and mistletoe do not.
d. The yellowhammer is from a southeastern state.
e. The willow ptarmigan is not from the camellia state.

State				
Flower				
Bird				

Which of the clues were the key(s) to solving the puzzle? Explain your reasoning.

Each year, the Calaveras County Frog Jumping Contest is held at Angel's Camp, California. In last year's contest, four large bullfrogs—Flying Freddie, Sailing Susie, Jumping Joe, and Leaping Liz—captured the first four places. Each frog was decorated with a brightly colored bow before the competition began. From the following clues, determine which frog won each place and the color of its bow.

a. Joe placed next to the frog with the purple bow.
b. The frog with the yellow bow won, and the frog with the purple bow was second.
c. The colors on Freddie's bow and Susie's bow mix to form orange.
d. The color of the remaining bow was green.

Construct a table similar to the one above to help organize your work.

Activity 14: Candies, Couples, and a Quiz

PURPOSE   Use the elimination strategy in problem-solving.

GROUPING   Work individually or in groups of 2 or 3.

GETTING STARTED   In the following problems, use some of the clues to construct a set of possible solutions. Then use the remaining clues to eliminate possibilities, or construct a table to help organize your work as in Activity 13.

Mike said to Linda, "Bet you can't guess the number of candies I have in this sack."

"Give me a clue," she said.

"I have more than 50 but fewer than 125. If you divide them into piles of eight, there are two left over. If you divide them into piles of seven, there is one left over," said Mike.

"Oh, that's easy, you have _____ candies!!" said Linda.

What is Linda's answer, and how did she get it so easily?

Four married couples are celebrating Thanksgiving together. The wives' names are Jolene, Jane, Marie, and Chris. Their husbands are Bob, Lyle, Lee, and Rick.

Examine the following clues and determine who is married to whom.

a. Rick is Jolene's brother.
b. Marie has two brothers, but her husband is an only child.
c. Lyle is married to Chris.
d. Jolene and Lee were once engaged but broke up when Lee met his present wife.

A social studies quiz consists of five true-false questions.

a. There are more true than false answers.
b. Questions one and five have opposite answers.
c. No three consecutive answers are the same.
d. Josh knows the correct answer to the second question.
e. From these clues, Josh can determine the correct answer to each question.

What are the correct answers to each of the five questions on the quiz?

Activity 15: What's the Last Card?

PURPOSE   Use problem-solving strategies and the four-step approach to solve a process problem.

MATERIALS   Fifteen index cards or one suit from a deck of playing cards

GROUPING   Work in groups of 2–4.

GETTING STARTED   Make a deck of nine cards numbered one through nine. Take the deck in one hand. Then place the cards on the table in the following manner.

- Place the top card on the table.

- Put the next card on the bottom of the deck.

- Place the third card on the table.

- Put the fourth one on the bottom of the deck.

- Continue this procedure until there is only one card left and place it on the table.

1. Was the last card you placed on the table the two? If not, put the cards back in order and try the experiment again.

2. a. If you start with 13 cards and follow the same procedure, which card do you think will be last? Why?

   b. Make a deck numbered one through 13 and try the experiment. What was the last card you placed on the table?

3. If you have a deck of 13 cards, it's easy to find the last card by going through the procedure. However, suppose you want to know the last card for a deck of 47 cards or one with 102 cards and you don't have that many cards. How can you predict the last card for a deck containing any number (*n*) of cards?

EXTENSIONS   In what order should the cards in a 15-card deck be arranged so that when they are placed on the table using the procedure above, the pile of cards on the table is in the correct numerical order one through 15?

Activity 16: Magic Number Tricks

PURPOSE   Translate verbal phrases into algebraic expressions.

MATERIALS   Calculator

GROUPING   Work individually.

Dr. Wonderful, the Mathematical Magician, astounds crowds with his amazing ability to read people's minds. Here are the directions that he gives to five people in the crowd. Follow Dr. Wonderful's directions and complete the table below. Choose a different number for each person.

PERSON	1	2	3	4	5
1. **Pick any number.**	___	___	___	___	___
2. **Multiply by 3.**	___	___	___	___	___
3. **Add 30.**	___	___	___	___	___
4. **Divide by 3.**	___	___	___	___	___
5. **Subtract your original number.**	___	___	___	___	___

Dr. Wonderful tells the people to write their answers on a sheet of paper but not to reveal them to anyone else. He closes his eyes, concentrates deeply, and then claims that the knows each person's answer.

Let n be the number. Write an algebraic expression and record the result for each step above.

Step 1:

Step 2:

Step 3:

Step 4:

Step 5:

The results is _____.

One of Dr. Wonderful's other mind-reading tricks involves birthdays. Use a calculator and follow along with the crowd as he gives the directions. Press ⟨=⟩ on your calculator after each step.

1. Enter the month of your birthday.

2. Multiply by 5.

3. Add 20.

4. Multiply by 4.

5. Subtract 7.

6. Multiply by 5.

7. Add the day of your birthday.

8. Subtract the number of days in a non-leap year.

"Oh," "Ah," and "Look at that" can be heard throughout the crowd. What do people see on the display of their calculator?

To the right of each step above, write an algebraic expression that correctly describes Dr. Wonderful's direction. Study the sequence of expressions and then explain how place value helps to explain how this "magic number trick" works.

EXTENSIONS   Write some magic tricks of your own. Write the algebraic expression for each step so that you can justify your final result. Try them out on your classmates.

Chapter 1 Summary

In this chapter, you studied some of the tools and processes used to solve problems. The goal was to help you develop your own problem-solving ability.

The study of patterns and functions is a central theme in mathematics. In this chapter, you learned various ways to analyze patterns. First you learned to recognize different types of patterns.

a. Patterns, such as those in Exercise 1 of Activity 1 and Exercise 1 of Activity 2, that have a repeating core:

AAB AAB AAB . . .   and   ABC ABC ABC . . .

b. Patterns, like those in Exercise 3 of Activity 1 and Exercise 2 of Activity 2, that have a growing core:

ABA ABBA ABBBA . . .   and   AB ABC ABCD . . .

c. Patterns that grow:

2, 5, 8, 11, . . .

•,   • •,   • • •,   . . .

d. Nested patterns, like those in Exercise 9 of Activity 2, that combine two or more patterns into one:

Pattern A	2,		5,		8,		11,		. . .
Pattern B		3,		5,		7,		9,	. . .
Nested pattern	2,	3,	5,	5,	8,	7,	11,	9,	. . .

While studying patterns, you learned some new terminology: *term*, *term number*, and *sequence*. You also learned that for many sequences, there is a rule (function) that relates the term of the sequence to the term number.

One method you learned for analyzing and extending numeric sequences was to examine the differences between the successive terms of the sequence. You found that in the cases where the differences are constant, you could use the difference to generate a rule for the general term of the sequence.

The methods you used to analyze and extend patterns based on your observations are examples of *inductive reasoning*. You discovered the limitations of the inductive reasoning process in Activity 4. No matter how many initial terms of a sequence you may know, there is generally more than one way to extend it.

Activities 5 and 6 used the *pattern* problem-solving strategy to introduce exponents and exponential growth.

In Activity 7, you explored a four-step approach to problem solving. One advantage of this approach is that it gives you a way to get started on solving a problem. Many of the remaining activities developed problem-solving strategies, such as *look for a pattern*, *simplify*, and *work backward*, which are often used in conjunction with the four-step approach.

Activities 9 and 16 integrated many problem-solving techniques. You learned to simulate problems that could not be experienced firsthand, to apply the pattern strategy, and to use tables as an organizer. You also learned a new problem-solving strategy, *simplify the problem—* begin with a simple case of the problem and work through successively more complex cases until a general method of solution is discovered. This technique will be used extensively for investigating new mathematical concepts.

The *work backward* problem-solving strategy was introduced in Activity 8 and the *logical reasoning* strategy was developed in Activities 10 - 14. Most of the activities began with a set of clues, or premises, that were accepted as true. By reasoning logically from these premises, you were able to conclude something about a number or situation. This process of deriving a conclusion by reasoning logically from a set of known premises is called *deductive reasoning*.

Usually, you were able to reason *directly* from the premises to the conclusion. However, in Activity 13, you had to test possible solutions by assuming they were true. If the assumption led to a contradiction of a known fact, then you knew that the proposed solution was not correct. This method, which was introduced through elimination, is known as *indirect* reasoning - it is used extensively in mathematics.

Logical reasoning is the cornerstone upon which mathematics is built. New mathematics is often discovered via inductive reasoning. But before a conjecture arrived at inductively is accepted as a fact, it must first be verified using deductive reasoning. It is this standard of proof that distinguishes mathematics from the other sciences.

You also engaged in the investigation of some classic number tricks in Activity 15. As you analyzed the tricks, you discovered how using variables to translate the instructions into algebraic expressions could help explain the "magic."

The activities in this chapter were intended to provide only an informal introduction to inductive and deductive reasoning. You will learn more about these techniques later in this book and use them throughout it.

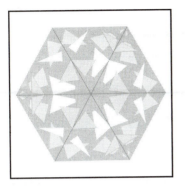

Chapter 2
Sets, Whole Numbers, and Functions

". . . understanding number and operations, developing number sense, and gaining fluency in arithmetic computation form the core of mathematics education for elementary grades. . . . As students gain understanding of numbers and how to represent them, they have a foundation for understanding relationships among numbers."
—Principles and Standards for School Mathematics

The activities in this chapter make connections among many of the NCTM Standards. Using the concepts associated with sets, identifying the properties of the elements of a set, and sorting the elements into different sets according to specific properties are all fundamental to the structure of mathematics.

Some of the activities in this chapter will engage you in sorting and classifying objects, describing their properties (attributes), and describing their similarities and differences. All of these activities promote communication in mathematics. Explaining your reasoning, defending your conjectures, and evaluating input from others will all promote your greater understanding of mathematical ideas.

When engaged in problem-solving situations that involve sorting, classifying, and discriminating, reflect on the connections between these processes in science and mathematics. You will find that the direct and indirect reasoning skills you develop in this chapter will be an important asset in other problem settings throughout the book.

The number activities in this chapter are designed to help you develop a strong sense of number, the concept of exponents, and an understanding of the basic operations of multiplication and division.

The concept of function is one of the central unifying themes in mathematics. The study of arithmetic, algebra, geometry, probability, and statistics relies on generalizing patterns and developing mathematical models (functions) that can be used to describe real world situations. The final activity in the chapter will extend the idea of describing a pattern by an informal *rule* to creating multiple representations of functions.

Activity 1: Attribute Elimination

PURPOSE	Use the elimination strategy to identify attribute pieces.
MATERIALS	Set of Attribute Pieces (page A-5)
GROUPING	Work individually or in pairs.

In Exercises one through four, use the clues to eliminate all but one attribute piece found on page A-5. Write the name of the piece in the blank.

1. It is yellow.
    It is square.
    It is not large.

    _____

2. It has three sides.
    It is not small.
    It is blue.

    _____

3. It is small.
    You spell its color with
      five letters.
    It does not have straight sides.

    _____

4. It is red.
    It is not a square.
    It is large.
    It is not a circle.
    It is not a triangle.

    _____

In Exercises one through three, a set of four attribute pieces is given. In each set, one piece is different from the others. For each set, decide which piece does not belong in the given set and explain your reasoning. If there is more than one possibility, explain.

1.

2.

3. (see figures)

Activity 2: Sort and Classify

PURPOSE   Reinforce the concept of an attribute through sorting and classifying attribute pieces.

MATERIALS   Set of Attribute Pieces (page A-5) and Attribute People Pieces (page A-1)

GROUPING   Work individually or in pairs.

---

1. Sort the attribute pieces according to size, shape, and color and complete the following table.

Sort the Pieces by	Number of Piles	Number in Each Pile
Size		
Color		
Shape		

2. How many attribute pieces are there? Explain how you can answer this question without counting the pieces one by one.

---

1. Sort the people pieces according to person, size, jacket color, and jacket style and complete the table.

Sort the People Pieces by	Number of Piles	Number in Each Pile
Person		
Size		
Jacket Color		
Jacket Style		

2. How many people pieces are there? Explain how you can answer this question without counting the pieces one by one.

Activity 3: Attribute Matrix

PURPOSE	Identify common attributes and complete a pattern on a matrix card.
MATERIALS	One set of Attribute Pieces (page A-5) and Matrix Cards (pages A-17–A-19)
GROUPING	Work individually or in pairs.
GETTING STARTED	Attribute pieces have been placed on each of the following matrix cards. A different pattern was used to place the pieces on each card. For each one, place the attribute pieces shown on a blank 3×3 Matrix Card. Then choose appropriate pieces from the remaining attribute pieces and place them in the blank cells to continue the pattern.

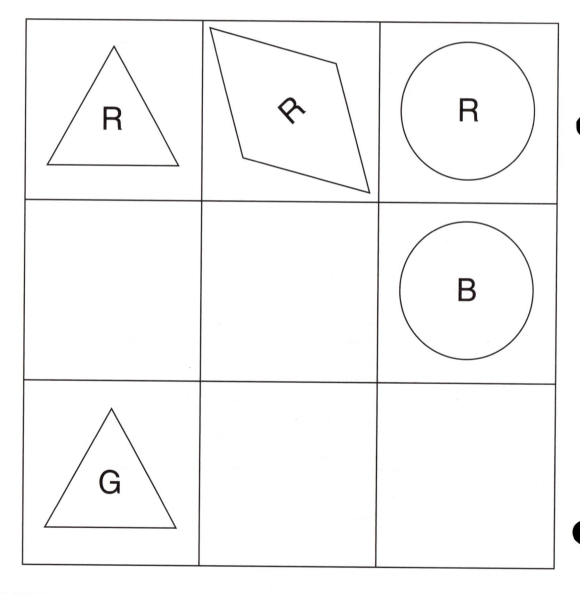

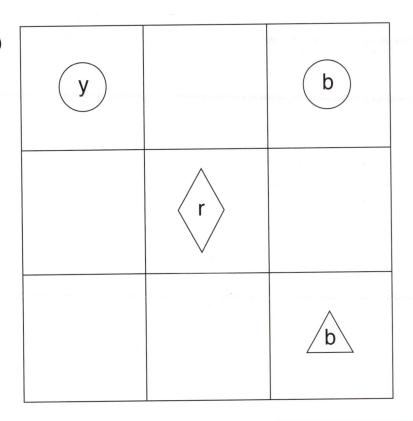

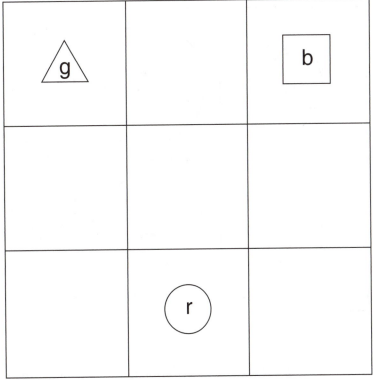

EXTENSIONS Use the blank 2 × 2, 3 × 3, and 4 × 4 matrix cards to make additional Attribute Matrix Puzzles. Begin a pattern on each matrix using as few attribute pieces as possible. Exchange puzzles with a classmate and complete each other's puzzles.

Activity 4: Difference Puzzles

PURPOSE   Determine attribute pieces that differ from each other by one, two, three, or four attributes.

MATERIALS   Set of Attribute Pieces (page A-5) or Attribute People Pieces (page A-1), a set of Attribute Difference Puzzles (pages A-20–A-21), and Matrix Cards (pages A-17–A-19)

GROUPING   Work individually or in pairs.

GETTING STARTED   Choose a matrix card and place attribute pieces in the cells so that each piece is different from the adjoining piece, horizontally and vertically, but not diagonally, by one attribute. Begin with the 2×2 matrix and progress to the 4×4 matrix. Then repeat the activity using two, three, and four differences on the 2×2, 3×3, and 4×4 matrix cards.

Example: 2×2 matrix with one difference

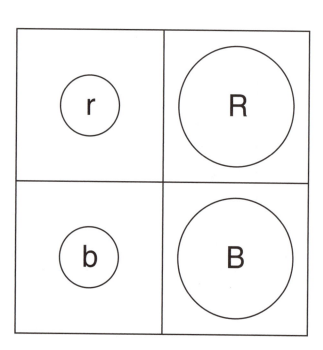

EXTENSIONS   Complete the Attribute Difference Puzzles on pages A-20–A-21 by placing the attribute pieces in the cells of each puzzle so that adjacent cells differ by the number of attributes shown in the diamonds between the cells. Use the blank matrix on page A-22 to make new Difference Puzzles for other students.

Activity 5: What's in the Loop?

PURPOSE	Use the elimination and logical reasoning problem-solving strategies to determine the identity of an attribute label card.
MATERIALS	One set of Attribute Pieces (page A-5) or Attribute People Pieces (page A-1), a large loop of yarn or string, and a set of Attribute Label Cards (pages A-23–A-24)
GROUPING	Work in pairs or in teams of two students each.
GETTING STARTED	Place the loop between the players. Shuffle the label cards and place them face down on the table. Player A picks a card from the pile, looks at it, and places it face down next to the loop without showing it to the other player. Player B chooses an attribute piece or a people piece and places it in the loop. Player A then tells whether the placement is correct or not based on the label card. Play continues until Player B can correctly identify the exact attribute on the label card. Players then switch roles.

Activity 6: Loop de Loops

PURPOSE	Explore the set concepts of union, intersection, and complement.
MATERIALS	One set of Attribute Pieces (page A-5), two loops of yarn or string, and a set of Attribute Label Cards (page A-23)
GROUPING	Work in pairs or in teams of two students each.
GETTING STARTED	Place the loops between the players. Put the **LARGE** label card face up on one loop and the **RED** label card face up on the other. Take turns placing pieces in the appropriate loop or outside the loops.

1. How must the loops be arranged so that all the attribute pieces can be placed correctly? Make a drawing to show the arrangement of the loops.

2. How many pieces are inside both loops, that is, how many are either RED or LARGE?

3. How many pieces are both RED and LARGE?

4. How many pieces are LARGE but not RED?

5. Repeat the activity with a different pair of label cards.

6. Repeat the activity using three loops and three label cards.

Activity 7: What's in the Loops?

PURPOSE   Use indirect reasoning to determine the attributes described on the label cards and reinforce union, intersection, complement, and subsets of a set.

MATERIALS   One set of Attribute Pieces (page A-5) or Attribute People Pieces (page A-1), two loops of yarn or string, and a set of Attribute Label Cards (pages A-23 and A-24)

GROUPING   Work in pairs or teams of two students each.

GETTING STARTED   Shuffle the label cards and place the deck between the players. Overlap the loops as shown below. Player A picks two label cards and, without showing them to the other player, places them face down, one on each loop as shown. Player B chooses an attribute piece and places it in one of the four regions. Player A then indicates whether the placement is correct or not according to the labels on the cards that have been placed on the loops. If the placement is incorrect, the first student may try another region, or put the piece back in the pile and try another attribute piece. Play continues until Player B can correctly identify both label cards. Players then switch roles.

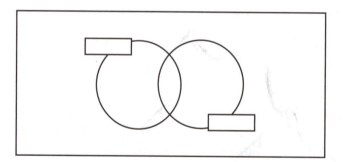

1. What labels could be used for the loops so that the intersection of the two loops is empty? (The sets are disjoint.)

2. What labels could be used for the two loops so that one loop would be contained in the other? (One set is a subset of the other.)

3. Repeat the activity using three loops and three label cards.

Activity 8: Odd and Even Patterns

PURPOSE To explore the sums of odd and even numbers and to discover the relationship between the two addends and the sum.

MATERIALS Scissors, one set of double-six dominoes, or cut out dominoes on page A-25, and shapes cut from centimeter graph paper (page A-49)

GROUPING Work in pairs.

GETTING STARTED Cut six sets of the number shapes shown below from grid paper and if necessary cut out a set of dominoes.

Domino Game

- Remove the dominoes that have a blank. Place the remaining dominoes face down and mix well.

- Place the number shapes between the players.

- Player A selects a domino and the number shapes that match and then puts them together so that the result is a $2 \times n$ rectangle or a figure that is as close to a $2 \times n$ rectangle as possible.

- Record each turn as shown.

Example Domino

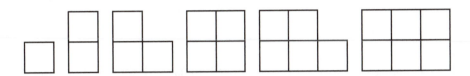

Number Shapes

Equation	2	+	5	=	7
Odd/Even	**Even**	**+**	**Odd**	**=**	**Odd**

- Players alternate turns. A player loses a turn when a pair of shapes cannot be found to match the domino.

- The game ends when neither player can make a match. The winner is the player with the greatest number of EVEN sums.

1. Explain why there is a greater number of even sums than odd sums. Use the result of putting the two shapes together to form the sum. Illustrate your answer with drawings.

2. Any even number can be written as ($2n$), where n is a whole number. Suppose a set of dominoes had up to 100 dots on one side. If you cut out shapes for the even numbers from grid paper, describe the even number shapes and how each shape is related to this algebraic expression.

3. If any even number can be written as ($2n$), then any odd number can be written as

   _____ or _____, where n is a whole number.

4. If you cut out shapes for odd numbers as previously described, describe the odd number shapes and how they relate to the algebraic expressions in the previous exercise.

5. Adding two even numbers can be represented as $2n + 2m = 2(n + m)$, where m and n are both whole numbers. Describe the shapes of the two addends and the sum as they relate to this algebraic equation.

6. Repeat Exercise 5 for the addition of two odd numbers and for the addition of an odd and an even number.

Activity 9: Multiplication Arrays

PURPOSE   Develop the concept of multiplication as repeated addition and the commutative property for multiplication.

MATERIALS   Colored Squares (page A-7), ceramic tiles, or squares cut from construction paper; square grid paper, a copy of the Multiplication and Division Frame (page A-26), and paper for recording

GROUPING   Work individually.

GETTING STARTED   In a multiplication problem, two factors determine a product. Every multiplication fact can be illustrated as a rectangular array on a multiplication and division frame. The factor to the left of the frame determines the number of groups (rows). The number above the frame determines the number in each group (columns).

Use the multiplication and division frame to construct each of the first 10 multiples of 4. Record your work on paper as shown below.

Example:

$$\begin{array}{c|c} & \text{factor} \\ \text{factor} & \text{product} \end{array}$$

Student Record Sheet

$1 \times 4 \rightarrow$ 1 group of 4 $\rightarrow$ 4

 1 $\times$ 4 = 4

$2 \times 4 \rightarrow$ 2 groups of 4 $\rightarrow$ 4 + 4

 2 $\times$ 4 = 8

$3 \times 4 \rightarrow$ 3 groups of 4 $\rightarrow$ 4 + 4 + 4

 3 $\times$ 4 = 12

---

1. What happens to the rectangular arrays and the products when the same factors are used but their order is reversed, that is 4×1, 4×2, 4×3?

2. On square grid paper, construct a 3×4 rectangle and a 4×3 rectangle. Cut out the rectangles and place one on top of the other. What must be done to align them so that they match each other? What can you conclude from this?

Activity 10: Find the Missing Factor

PURPOSE   Develop the concept of division as equal sharing (*partitioning*).

MATERIALS   Colored Squares (page A-7), a copy of the Multiplication and Division Frame (page A-26), and one die

GROUPING   Work individually or in pairs.

GETTING STARTED   The problem, "Tasha has 15 marbles. She wishes to give the same number of marbles to each of five friends. How many marbles does each person receive?" is an example of the *partitioning* model for division. Use the multiplication/division frame to solve $15 \div 5$ or $5)\overline{15}$. Record the factor (divisor) 5 to the left of the frame. Using 15 squares, determine how many squares must be placed in each of the five groups to form a rectangular array.

Example: a.   Place a square in each group.

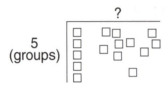

b.   Now arrange the rest of the squares in groups to construct a rectangular array.

c.   How many squares are there in each group (row)? _____

$$15 \div 5 = 3$$

Student one rolls a die to determine a factor. Write the number to the left of the frame. Student two picks a handful of squares (without counting) to determine the product (dividend), and then uses the factor to construct a rectangular array as in the previous example.

Example: The number of squares in the handful equals 15.

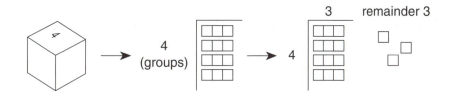

1. a. How many squares are in each of the 4 groups in the example? _____

    b. How many squares are remaining? _____

    c. Record your answer. $15 \div 4 =$ _____ R _____

2. Repeat this activity 10 times, alternating turns.

3. Describe the relationship between the *partitioning* model for division and multiplication.

EXTENSIONS   Use the *integer division* key on your calculator to solve the following problems:

   1. $64 \div 3$ 2. $42 \div 2$ 3. $116 \div 6$

   4. $57 \div 4$ 5. $94 \div 6$ 6. $43 \div 8$

   7. $81 \div 9$ 8. $79 \div 5$ 9. $157 \div 7$

   10. If you use the $\div$ key, $15 \div 4 = 3.75$. How would you use the calculator to determine the *whole number* remainder? Show the sequence of keys to be pressed.

Activity 11: How Many Cookies?

PURPOSE	Develop the concept of division as repeated subtraction.
MATERIALS	Colored Squares (page A-7) and paper for recording
GROUPING	Work individually or in pairs.
GETTING STARTED	The following is an example of the repeated subtraction model of division: "Tyler has 24 candies in a jar and wishes to decorate each cookie with six candies. How many cookies will he be able to decorate?" Solving $24 \div 6$ or $6\overline{)24}$, is illustrated below.

Example:

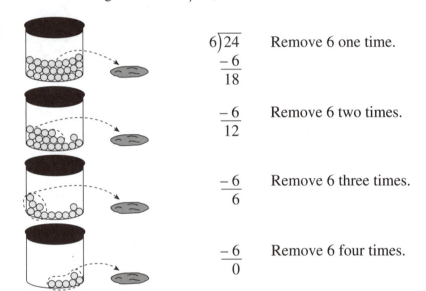

$$6\overline{)24}$$
$$\underline{-\,6}$$
$$18$$
Remove 6 one time.

$$\underline{-\,6}$$
$$12$$
Remove 6 two times.

$$\underline{-\,6}$$
$$6$$
Remove 6 three times.

$$\underline{-\,6}$$
$$0$$
Remove 6 four times.

$$24 \div 6 = 4$$

Write five additional word problems that illustrate the repeated subtraction model for division and solve them as shown above.

Solve the following problems using the *constant* feature of your calculator.

Example: $120 \div 20$

or

$$\boxed{ON}\ \boxed{\div}\ \boxed{2}\ \boxed{0}\ \boxed{Cons}\ \boxed{1}\ \boxed{2}\ \boxed{0}\ \boxed{Cons}\ \boxed{Cons}\ \boxed{Cons}\ \boxed{Cons}\ \boxed{Cons}\ \boxed{Cons}$$

$$\boxed{ON}\ \boxed{1}\ \boxed{2}\ \boxed{0}\ \boxed{\div}\ \boxed{2}\ \boxed{0}\ \boxed{=}\ \boxed{=}\ \boxed{=}\ \boxed{=}\ \boxed{=}\ \boxed{=}\ \boxed{=}$$

1. $96 \div 6$	2. $318 \div 3$	3. $107 \div 24$	4. $858 \div 13$
5. $94 \div 23$	6. $152 \div 4$	7. $297 \div 3$	8. $432 \div 17$

Explain how you determined the quotient and the remainder in each problem using the *repeated subtraction* model.

Activity 12: What's My Function?

PURPOSE	Determine the function relating input and output values.
MATERIALS	One set of Function Cards (pages A-27 and A-28).
GROUPING	Work in pairs.
GETTING STARTED	Players alternate turns trying to guess the function on one of the cards.

A player scores five points if the function is guessed on the first try, three points if it is guessed on the second try, and one point if it is guessed on the third try. Once the function is determined, one additional point is scored for each correct way a player can express the function: table, graph, equation, function, or verbal expression. The player who has scored the most points after all of the cards have been used is the winner.

To start the game, shuffle the cards and place them face down on the table. One player draws a card from the top of the deck without showing it to the other player. The other player tries to guess the function on the card by giving input values one at a time. The player holding the card applies the function on the card to the input value and responds with the output.

The guesser can give at most 10 input values and can guess the rule at any time. However, only three guesses are allowed per card.

Example:

What's My Function?

Multiply input by 2 and add 1.

$y = 2x + 1$ $f(x) = 2x + 1$

x	–1	0	1	2	3
y	–1	1	3	5	7

Input	Output	Guess	Response
2	5	Add 3 to the input.	Incorrect.
0	1	(No Guess)	
4	9	Multiply input by 2 and add 1. $y = 2x + 1$ $f(x) = 2x + 1$	Correct.

Score:

Guessed function on second try:	3 points
Expressed function 3 ways	3 points
Total	**6 points**

---

1. What strategies for choosing input values might make it easier to guess the function?

EXTENSIONS   Make up Function Cards of your own and use them to play the game.

Chapter 2 Summary

Sets and the operations on them were introduced in this chapter. The fields of mathematics, such as arithmetic, geometry, and algebra, are characterized by the study of particular sets of objects. This study entails an analysis of the elements of the set, their attributes, and what makes the set uniquely different from other sets.

In Activity 2, you sorted elements of a set according to certain attributes. Sorting and classifying were visited again with increasing complexity in subsequent activities. These activities develop an informal understanding of how precise definitions of terms—so important in mathematics—are generated.

The Fundamental Counting Principle was introduced in Activity 2. If we know the number of each of the attributes, that is, two sizes, three colors, four shapes, then the total number of pieces is the product of the numbers of the attributes: $2 \times 3 \times 4 = 24$. The Fundamental Counting Principle is important in probability and will be explored in more depth in Chapter 7.

Activity 2, which involves sorting and classifying the attribute and people pieces, and Activities 3 and 4, which deal with identifying differences and similarities among the pieces, introduce skills that are basic to the study of science as well as mathematics. For example, while mathematicians classify geometric shapes according to attributes, scientists sort all living creatures into a hierarchy of classifications. Each class differs from the previous one by some attribute that makes it unique. It is important to recognize this connection between the fundamental processes of mathematics and science and how activities such as these can help in developing basic understanding in both subjects in the elementary grades.

In Activities 5–7, you encountered the set concepts of *union*, *intersection*, *complement*, and the *empty set*. The set of all pieces provided an example of a *universal* set, the word OR described the *union* of sets, AND, the *intersection* of sets, and NOT, the *complement*. These concepts were illustrated by the following diagrams.

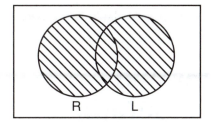

RED or LARGE
Union

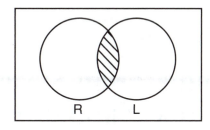

RED and LARGE
Intersection

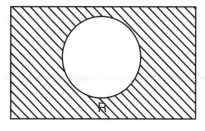

Not RED
Complement

In Activity 5, when you made a choice to place a particular attribute piece in the loop, you guessed (made an assumption) that a certain label belonged to that loop. Given a YES answer, you gained information about the correct label. Given a NO answer, you also gained information about which label was NOT correct, thus eliminating some labels, and reducing the number of pieces to try in order to complete the problem. Activity 7 extended this use of the elimination problem-solving strategy and the use of indirect reasoning to determine the proper label for the sets.

Activity 8 introduced the combination of *odd* and *even* numbers. After exploring the patterns generated by this addition, you determined an algebraic expression to generalize the addition.

In Activities 9–11, you were introduced to the basic concepts of multiplication and division of whole numbers. Activity 12 built on the lessons learned in Chapter 1. The mathematical concept of function was introduced through an activity involving patterns and a game using the guess and check strategy applied to the input and output values of various functions. The concept of function will be revisited in many activities throughout the book.

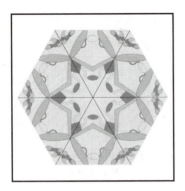

Chapter 3
Numeration Systems and Whole Number Computation

Give a man a fish,
and he eats for a day.
Teach him how to fish,
and he eats for a lifetime.

". . . students should attain rich understanding of numbers—what they are; how they are represented with objects, numerals or on number lines; how they are related to one another; how numbers are embedded in systems that have structures and properties; and how to use numbers and operations to solve problems. . . . Representing numbers with various physical materials should be a major part of mathematics instruction in the elementary school grades."
—Principles and Standards for School Mathematics

The development of number sense, operation sense, place value concepts, and an understanding of the algorithms for the basic operations are among the most important tasks in teaching elementary mathematics. In Chapter 2, you began developing number sense ideas and refined your understanding of operations. This chapter continues the emphasis on these concepts and makes connections among concrete models, numbers, and algorithms for the basic operations. Through these models and representations, the elementary teacher promotes greater conceptual development of number and operation sense.

This conceptual approach to teaching mathematics allows students to construct their own ideas of number and computation. The use of manipulatives provides a problem-solving environment for learning in which students are constantly discussing, conjecturing, asking new questions as they review their work, and communicating with others about their findings.

The activities in this chapter provide interesting and challenging situations to increase your number sense and your understanding of place value and algorithms for operations. Each one provides an opportunity to explore mathematical concepts in a hands-on, problem-solving setting or in a game format.

Activity 1: A Visit to Fouria

PURPOSE	Use a game to reinforce place-value concepts, to introduce the base-four numeration system, and to develop understanding of the regrouping process.
MATERIALS	Blue, red, and white chips (10 of each color) and a die
GROUPING	Work in pairs.
GETTING STARTED	While on an Intergalactic Numismatics Tour you encounter a meteor shower and are forced to make an unscheduled stop on the planet Fouria. The monetary system used on Fouria consists of three coins: a white coin (worth $1 in U.S. money), a red coin, and a blue coin. The red coin is equivalent in value to four white coins, and the blue coin is equivalent to four red coins.

1. Unlike its sister planet, Ufouria, Fouria turns out to be a rather dull place to visit. To help pass the time in the waiting area, you and a fellow passenger play a coin trading game. The rules of the game are:

   - Players alternate turns.

   - On each turn, a player rolls the die and places that number of white Fourian coins in the white column on his or her Coin Trading Game Sheet.

   - Whenever possible, a player must trade four white coins for one red coin and/or four red coins for one blue coin.

   - Coins must be placed in the appropriately labeled column. No more than three coins of one color may be in any column at the end of a turn.

   - The first player to get two blue coins is the winner.

Make a Coin Trading Game Sheet, and play the game with a partner. At the end of each turn, record the number of each color coin on your game sheet in a table like the one at the right.

Coin Trading Game Sheet		
Blue	**Red**	**White**

Turn	Number Rolled	Result		
		B	**R**	**W**
1				
2				
3				
4				
5				
6				
7				
8				
9				
10				

The Coin Trading Game Revisited

1. A third passenger has been watching you play. She suggests it is more challenging to start the game with three blue coins and to remove the number of white coins equal to the number rolled on each turn. The first player to remove all the coins from his or her playing board is the winner. Your playing partner is confused. "How can you remove white coins when there aren't any on the board?" he asks. Explain how this can be done.

2. Play this new version of the game with a partner. Each player should record the result of each move in a table like the one on the preceding page.

---

1. You become bored with the games and go to the spaceport newsstand to buy something to read. Glancing at the cover of a Fourian magazine you notice that the price is given as 123_4. At the checkout stand, the clerk explains that this means one blue coin, two red coins, and three white coins. How many of each color coin does each of the following prices represent?

    a. 231_4 _____

    b. 102_4 _____

    c. 13_4 _____

    d. 20_4 _____

2. How would Fourians write each of the following prices?

    a. 1 blue, 2 red, 1 white _____

    b. 2 red, 3 white _____

    c. 2 blue _____

    d. 2 blue, 3 white _____

3. Back in the waiting area, you begin leafing through a magazine that you purchased. You note that the first page is numbered 1_4, but when you get to the fourth page, you are surprised to find that it is numbered 10_4. Fill in the blanks below to show how the remaining pages of the magazine would be numbered.

    1_4 ____ ____ 10_4 ____ ____ ____ ____ ____

    ____ ____ ____ ____ 33_4 ____ ____ ____ ____ ____

    ____ ____ ____ ____ ____ ____ ____ ____ ...

    ____ ____ 333_4 ____ ____ ____ ____ ____ ____

1. After reading for a while, you decide to have dinner. The price of your meal was 121_4. When you go to the cashier to pay for your meal, you realize that you don't have any Fourian money with you. "No problem," the cashier explains. "You may pay with dollars." What is the cost of your meal in dollars?

2. a. On your way back to the waiting area, you stop in the newsstand to buy a souvenir. It costs 1312_4. How many dollars is this?

   b. You give the cashier two $100 bills. How much Fourian money should you get back in change?

---

1. Back in the waiting area, you find a mathematics book left behind by a Fourian student. Flipping through the book, you come across the examples shown below. Explain what the small 1 in the second example means and how it was obtained.

$$\begin{array}{r} 2\,3\,1_4 \\ +\,1\,0\,2_4 \\ \hline 3\,3\,3_4 \end{array} \qquad \begin{array}{r} {}^{1} \\ 2\,3_4 \\ +\,3\,3_4 \\ \hline 1\,2\,2_4 \end{array}$$

2. Use what you learned in the examples in Exercise 1 to find the following sums.

   a. $121_4 + 211_4$ b. $123_4 + 221_4$

3. A few pages later, you find the following examples. Explain what is being done in steps A, B, and C.

$$\begin{array}{r} 3\,3\,3_4 \\ -\,1\,2\,1_4 \\ \hline 2\,1\,2_4 \end{array} \qquad \begin{array}{r} {}^{2}\cancel{3}\,{}^{1}1_4 \leftarrow \text{(A)} \\ -\,1\,2_4 \\ \hline 1\,3_4 \end{array}$$

   (C) (B)

4. Use what you learned in the examples in Exercise 3 to find the following differences.

   a. $323_4 - 211_4$ b. $221_4 - 132_4$

EXTENSIONS

1. Explain how the place-value system used to write Fourian numerals is related to the place-value system used to write base ten numerals.

2. Explain how the regrouping used in Fourian addition and subtraction is related to the regrouping used in regular (base ten) addition and subtraction.

Activity 2: Regrouping Numbers

PURPOSE   Develop number sense and an understanding of place value and regrouping concepts by constructing models of numbers.

MATERIALS   Hundreds, tens, and ones place-value dice (see page A-30), one Place-Value Operations Board (page A-29), and Base Ten Blocks (pages A-13 and A-15).

GROUPING   Work individually or in pairs.

GETTING STARTED   Begin with the tens and ones dice. One student rolls the dice; the second student uses rods and cubes to construct the two-digit number represented by the dice on the place-value operations board. The number should be constructed first using the least number of blocks possible. One of the highest place-value blocks should then be traded for 10 of the next lower order blocks. Each number should be recorded as shown in the example. The trading should continue until only ones cubes are used. Students then switch roles.

Example:

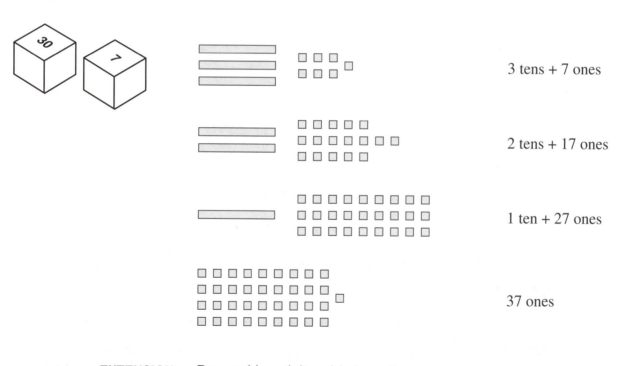

3 tens + 7 ones

2 tens + 17 ones

1 ten + 27 ones

37 ones

EXTENSION   Repeat this activity with three dice.

Complete the chart.

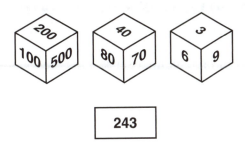

243

▨	▬	▫
2 hundreds	4 tens	3 units
1 hundreds	__ tens	3 units
1 hundreds	12 tens	__ units
2 hundreds	3 tens	__ units
0 hundreds	23 tens	__ units
__ hundreds	12 tens	23 units

Explain why rewriting a number in several different ways leads to an understanding of the regrouping (renaming) process used in computational algorithms.

EXTENSIONS   Duplicate several charts like the one shown below. Fill in some of the parts and then have your partner complete the chart.

[ ]

▨	▬	▫
__ hundreds	__ tens	__ units
__ hundreds	__ tens	__ units
__ hundreds	__ tens	__ units
__ hundreds	__ tens	__ units
__ hundreds	__ tens	__ units
__ hundreds	__ tens	__ units

Activity 3: Find the Missing Numbers

PURPOSE Reinforce number sense and place-value concepts in a problem-solving situation.

MATERIALS Base Ten Blocks (pages A-13 and A-15) and Place-Value Operations Board (page A-29)

GROUPING Work in pairs.

GETTING STARTED One student reads the clues to the other. The second student uses blocks to construct a model of the missing number(s). Students alternate turns.

---

1. I have five base ten blocks. Some are rods and some are units. Their value is less than 20.

    Who am I? _____

---

2. I have eight base ten blocks. Some are units and some are rods. Their value is an odd number between 50 and 60.

    Who am I? _____

---

3. I have four base ten blocks. Some are rods and some are units. Their value is between 30 and 40.

    Who am I? _____

---

4. I have six base ten blocks. Some are flats, some are rods, and some are units. I am a palindrome.

    Who am I? _____

---

5. I have two base ten blocks.

    Who am I? _____

---

6. I have three base ten blocks. None of them are flats.

    Who am I? _____

---

7. I have four base ten blocks. None of them are flats.

    Who am I? _____

---

8. I have four base ten blocks. Only one of them is a flat.

    Who am I? _____

EXTENSIONS Create additional clue cards for your partner to solve.

Activity 4: It All Adds Up

PURPOSE   Develop an understanding of addition of multidigit numbers and regrouping.

MATERIALS   Place-Value Operations Board (page A-29), Base Ten Blocks (pages A-13 and A-15), and tens and ones place-value dice (see page A-30).

GROUPING   Work in pairs.

GETTING STARTED   The first student rolls the dice and constructs the two-digit number on a place-value operations board using the correct base ten blocks as in the example. The number should also be recorded on a student record sheet. Then the second student rolls the dice and constructs his or her number beneath that of the first student as shown. The number is also recorded below the first one on the record sheet and a line is drawn across the sheet below the two numbers. The blocks are then moved together to represent the sum of the two numbers. If a column has 10 or more blocks, then a 10-for-1 trade should be made. Record the representation of the sum.

Example: **Place-Value Operations Board**   **Student Record Sheet**

Student 1

2 tens + 8 ones   28

Student 2

3 tens + 4 ones   + 34

5 tens + 12 ones

Make 10-for-1 trade

6 tens + 2 ones =   62

EXTENSION   1. Roll the dice to make five additional addition problems. Find the sum as shown above.

            2. You are probably familiar with the use of the term *carry* to describe part of the addition algorithm. Explain how *regrouping* or *renaming* more accurately describes this process than the word *carry*.

Activity 5: What's the Difference?

PURPOSE   Use the *take-away* and *comparison* models to develop an understanding of subtraction of multidigit numbers.

MATERIALS   Place-Value Operations Board (page A-29), Base Ten Blocks (pages A-13 and A-15), and 2 sets of place-value dice (tens and ones).

GROUPING   Work in pairs.

GETTING STARTED   For the *take-away* model of subtraction, each student rolls a set of dice. The person with the larger number constructs it on the operations board using base ten blocks. The other student removes (*takes away*) the number of blocks represented by the two-digit number on his or her dice. Record the representation of the difference.

Example 1:

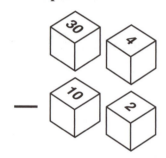

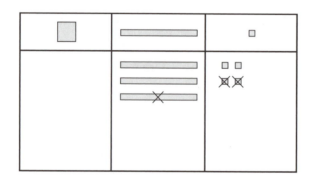

$$\begin{array}{r} 34 \\ -12 \\ \hline 22 \end{array}$$

Example 2:   If necessary, make a 1-for-10 trade.

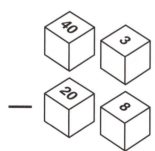

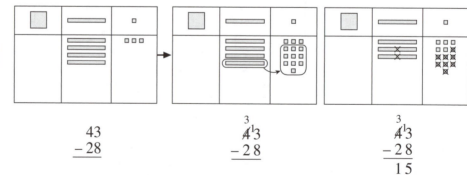

Roll the dice to make five additional subtraction problems. Determine the difference between each pair of numbers using the *take-away* model. Record each subtraction problem and the solution. Show *regrouping* if necessary.

For the *comparison* model of subtraction, each student rolls a set of dice. The person with the larger number constructs it at the top of the place-value operations board using base ten blocks. The other student constructs the smaller number below the first, as shown in Example 3. In order to make the comparison, stack the blocks representing the smaller number on top of those representing the larger number. If necessary, make a 1-for-10 trade. The blocks that are not covered represent the difference between the two numbers.

Example 3:

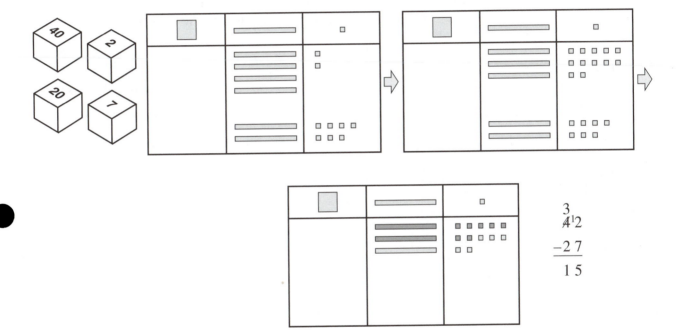

$$\begin{array}{r} \overset{3}{\cancel{4}}{}^{1}2 \\ -27 \\ \hline 15 \end{array}$$

1. Roll the dice to make five additional subtraction problems. Determine the difference between each pair of numbers using the *comparison* model. Record each subtraction problem and the solution. Show *regrouping* if necessary.

2. Write two word problems for the *take-away* model for subtraction, and two for the *comparison* model. Each problem should be a real world application relevant to the life of an elementary student and correctly represent the designated model.

3. Explain the role that *regrouping (renaming)* of numbers plays in multidigit addition and subtraction.

Activity 6: Multidigit Multiplication

PURPOSE   Develop the multiplication algorithm with multi-digit numbers and reinforce place-value concepts.

MATERIALS   Base Ten Blocks, place-value dice (tens and ones), and Multiplication and Division Frame (page A-26).

GROUPING   Work individually or in pairs.

GETTING STARTED   Roll the dice to generate two-digit numbers as shown. Construct these numbers on the top and left of the frame using blocks. Then construct a rectangle inside the frame with the appropriate base ten blocks.

Example:

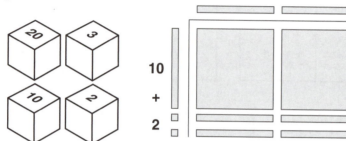

Count the blocks to determine the product.

2 flats   = 200

7 rods   = 70

6 units   =   6

Product = 276

Record your work in expanded notation to reinforce place value in the factors and in the product.

$$
\begin{array}{rrrrrrr}
 & & & 20 & + & 3 & \\
 & & \times & 10 & + & 2 & \\
\hline
 & & & 40 & + & 6 & \\
200 & + & 30 & & & & \\
\hline
200 & + & 70 & + & 6 & = & 276 \\
\end{array}
$$

Roll the place-value dice to generate factors for five additional multiplication problems. Solve the problems using a multiplication and division frame as illustrated in the example above. Record all answers in expanded form.

EXTENSIONS   Use a multiplication and division frame and base ten blocks to illustrate multiplication of binomials. Let a flat equal x^2, a rod equal x, and a unit cube equal 1.

Example: $x + 2 = $ ⬭⬭⬭⬭ + ☐ ☐

Complete the following:

1. $(x + 2)(x + 3) =$ 2. $(x + 4)(x + 5) =$

3. $(x + 1)(x + 7) =$ 4. $(x + 6)(x + 8) =$

Activity 7: Multidigit Division

PURPOSE Develop the division algorithm and reinforce the *partitioning model* and the place-value concepts associated with division.

MATERIALS Base Ten Blocks (pages A-13 and A-15), three place-value dice (hundreds, tens, and ones), Place-Value Operations Board (page A-29), and paper squares approximately 15 cm square.

GROUPING Work individually or in pairs.

GETTING STARTED Roll the place-value dice to generate a three-digit product (the dividend) and use base ten blocks to construct it on a place-value operations board. Roll the units die to determine the factor (divisor). The divisor determines the number of paper squares needed to represent the number of groups.

Example: 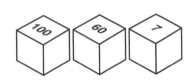 $5\overline{)167}$

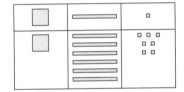

 → number of groups

Solution:

a. Place an equal number of the highest place-value blocks (flats) on each square. If this is not possible, record a 0 in the hundreds place above the division frame as shown below. Make a 1-for-10 trade with the next lower place-value block (rods) as shown at the right. In some cases, it may be necessary to make more than one trade.

$$\frac{0}{5\overline{)167}}$$

b. There are now 16 rods. Place an equal number of rods on each square. The rods placed on one of the squares is equal to what whole number? _____

Record these results as follows:

$$\begin{array}{r} 030 \\ 5\overline{)167} \\ \text{tens} \\ 150 \\ \hline 17 \end{array}$$

c. Proceed as shown on the previous page. Make a 1-for-10 trade and share an equal number of ones blocks among the squares. Record the results of the sharing as shown.

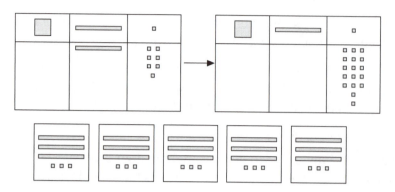

$$\begin{array}{r} 3 \\ 030 \\ 5\overline{)167} \\ 150 \\ \hline 17 \\ \text{units} \\ 15 \\ \hline \end{array}$$

The units placed on one of the squares is equal to what whole number? _____

How many blocks remain? _____ Can they be shared equally among the squares? _____

Record the final results as follows:

When 167 is distributed among 5 groups, there are 33 in each group and 2 left over that cannot be shared (the remainder).

$$\begin{array}{r} 3 \\ 030 \\ 5\overline{)167} \\ 150 \\ \hline 17 \\ 15 \\ \hline 2 \end{array} = 33 \text{ remainder } 2$$

Use the place-value dice to generate a factor (divisor) and a product (dividend) for six problems. Use base-ten blocks and the method described above to model the division. Record results as illustrated in the example.

EXTENSIONS   Explain the role that *regrouping* (*renaming*) of numbers plays in multiplication and division with multi-digit numbers.

Activity 8: Division Array

PURPOSE   Use the array model to demonstrate division as the inverse operation
of multiplication.

MATERIALS   Base Ten Blocks (pages A-13 and A-15), three place-value dice, the
Multiplication and Division Frame (page A-26), and a Place-Value
Operations Board (page A-29)

GROUPING   Work individually.

GETTING STARTED   Roll the three dice to generate a three-digit number, the product
(dividend). Use base ten blocks to construct the number on a
place-value operations board.

Example: 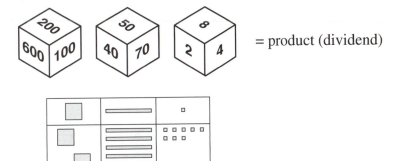   = product (dividend)

Roll the ones die to generate one factor (divisor) and place it outside
the multiplication-division frame as shown. This factor determines
how many groups (or rows) need to be constructed. The goal is to
distribute all blocks of the product evenly to form a rectangular array.

Example:

Since the factor (divisor) is 6 and there are only 2 flats, they cannot
be distributed evenly into the 6 rows. Make a 1-for-10 trade for each
flat and construct an array as shown by sharing the 25 rods as evenly
as possible.

Missing Factor

Next, make a 1-for-10 trade for the remaining rod.

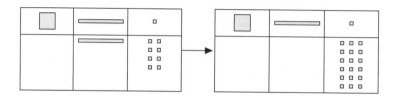

Now, distribute the 18 units evenly among the 6 rows.

Missing Factor

The number represented by the rods and units in each row (**43**) is the missing factor (quotient).

Generate a factor (divisor) and a product (dividend) for six division problems using the place-value dice. Use base ten blocks, a multiplication-division frame, and the method shown above for multidigit division to demonstrate the array model for division.

EXTENSION   Compare the work in this activity to that in Activities 9 and 10 in Chapter 2. Explain how this model reinforces the understanding of the inverse relationship between division and multiplication.

Activity 9: Target Number

PURPOSE Apply the guess and check problem-solving strategy to reinforce the inverse relationship between multiplication and division, and develop estimation skills.

MATERIALS One calculator for each pair of students

GROUPING Work in pairs.

GETTING STARTED Construct several tables like the one shown below. Only whole numbers may be used in the games. The calculator may only be used to find the product or quotient as indicated in each game. All other factors must be determined mentally.

TARGET = _____			
Turn	Factors	Product	Difference
1			
2			
3			
		TOTAL	

GAME 1

- Players choose a target number, and each player records it in a table.

- On alternate turns, each player chooses two factors (other than 1 and the target) and records them in the table. Then the player uses the calculator to multiply the factors and records the product.

- If the product equals the target, the player wins. If not, the player records the absolute value of the difference between the product and the target number in the table.

- If neither player has won after three turns, the person with the least total for the three differences is the winner.

GAME 2

- Player 1 chooses a target number; Player 2 chooses a constant factor and an acceptable range for the answer, for example ±20.
- On alternate turns, each player chooses another factor and uses the calculator to multiply it by the constant factor to obtain a product.
- If the absolute value of the difference between the product and the target number is within the range chosen by the players, the player wins.
- If the absolute value of the difference is outside the range, the players alternate turns until one player obtains a product within the specified range of the target number

SAMPLE GAME

Target Product = 650   Constant Factor = 19   Range = ±8

	Constant Factor		Estimated Factor		Product	| Difference |
Player 1	19	×	37	=	703	53
Player 2	19	×	30	=	570	80
Player 1	19	×	33	=	627	23
Player 2	19	×	34	=	646	4

Player 2 wins!

EXTENSIONS

1. Repeat **GAME 1** using division. On each turn, the player chooses a dividend and a divisor and divides them. If the quotient equals the target number, the player wins. If not, the player records the difference between the quotient (rounded to the nearest whole number) and the target in the table.

2. Repeat **GAME 2** using division. Player 1 chooses the target quotient. Player 2 chooses the constant divisor and an acceptable range for the answer.

SAMPLE GAME 2

Target Quotient = 26   Constant Divisor = 14   Range = ±1

	Estimated Dividend		Constant Divisor		Quotient	| Difference |
Player 1	320	÷	14	≈	22.9	3.1
Player 2	380	÷	14	≈	27.1	1.1
Player 1	365	÷	14	≈	26.07	0.07

Player 1 wins!

Activity 10: Least and Greatest

PURPOSE Use the guess and check problem-solving strategy to develop number and operation sense and to reinforce the concept of place value and the distributive property.

MATERIALS A calculator.

GROUPING Work individually.

---

1. Arrange the digits 1, 2, 3, 4, and 5 to make a two-digit and a three-digit number. Each digit may be used only once. Use a calculator to multiply the numbers. Try several different arrangements of the digits to determine the arrangement that results in the **greatest** possible product.

   __ __ __   __ __ __   __ __ __
   × __ __   × __ __   × __ __
   _____   _____   _____

2. Repeat the problem and arrange the digits so that you obtain the **least** possible product.

   __ __ __   __ __ __   __ __ __
   × __ __   × __ __   × __ __
   _____   _____   _____

3. Analyze the results in Exercises 1 and 2. Given any five non-zero digits $a < b < c < d < e$, what placement of the digits in a two-digit and a three-digit number will guarantee the **greatest** or the **least** product? Explain why you placed the digits in each place-value position.

EXTENSIONS Given any seven different non-zero digits, what placement of the digits in a two-digit number and a five-digit number, or a four-digit and a three-digit number, will guarantee the **greatest** and the **least** product?

Does the arrangement of five or seven digits also work for arranging six digits in two three-digit numbers to guarantee the **greatest** product? Explain.

Chapter 3 Summary

"Number sense refers to an intuitive feeling for numbers and their uses and interpretations; an appreciation for various levels of accuracy when figuring; the ability to determine arithmetical errors; and a common sense approach to using numbers."

—*Developing Number Sense*
Addenda Series, Grades 5–8

The development of number sense, operation sense, and a basic understanding of algorithms for operations is a focal point of the mathematics curriculum described in the *Principles and Standards for School Mathematics*.

The first activities in this chapter are examples of a curriculum design that begins with concrete models (base ten blocks), and proceeds through the representational (connecting models and symbolism) to the abstract (numeral). Activity 3 reinforces the understanding of our base ten numeration system through the investigation of a base four numeration system.

Activities 4 and 5 develop the algorithms for addition and subtraction of whole numbers. Numbers were modeled on the Place-Value Operations Board; 10-for-1 or 1-for-10 trading demonstrated the regrouping necessary for addition and subtraction, respectively. Recording each step of the activity on the student record sheet helped develop a clear understanding of the addition and subtraction algorithms as well as the associated place-value concepts.

The *comparison* model for subtraction was emphasized since it is the best way to explain the concepts of "how many more (greater) than?" and "how many less (fewer) than?" The model also illustrated the concept of subtraction as the inverse of addition. When the blocks representing the number to be subtracted were placed on top of the blocks of the larger number (see example in Activity 5), the remaining blocks represented the number that must be added to the smaller number to obtain the larger. This concrete comparison of the two numbers promotes an understanding of the quantity of one number in relation to another.

Activities in Chapter 2 developed an understanding of multiplication and division and also illustrated the concept of inverse operations. In this chapter, this basic understanding was extended to the use of the algorithms with larger numbers. The two words *factor* and *product* were stressed in both of these operations rather than the traditional *multiplier*, *multiplicand*, *dividend*, *divisor*, and *quotient*. This focus helps to reinforce the inverse relationship between the operations.

In the final two activities, the calculator was used in a problem-solving setting to reinforce all the principles of the earlier activities. Investigation of the partial products obtained from multiplying the various arrangements of the digits in Activity 10 developed a deeper understanding of place value concepts and lead to the arrangement of digits needed to obtain the largest or smallest product. Estimation and mental arithmetic strategies are important features of the Target Number activity.

The relationship among the four basic operations are illustrated in two different diagrams below. Notice the connections between multiplication as repeated addition and division as repeated subtraction.

	Once	Many
Join	+	×
Separate	−	÷

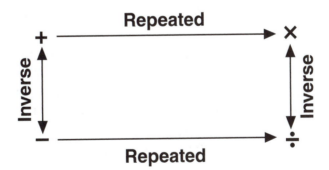

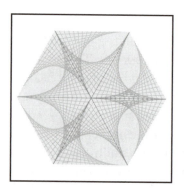

Chapter 4
Integers and Number Theory

"In lower grades, students may have connected negative numbers in appropriate ways to informal knowledge derived from everyday experiences, such as below-zero winter temperatures or lost yards on football plays. In the middle grades, students should extend these initial understandings of integers. Positive and negative integers should be seen as useful for noting relative changes or values. Students can also appreciate the utility of negative integers when they work with equations whose solution requires them, such as $2x + 7 = 1$."
—*Principles and Standards for School Mathematics*

"Number theory offers many rich opportunities for explorations that are interesting, enjoyable, and useful. These explorations have payoffs in problem-solving, in understanding and developing other mathematical concepts, in illustrating the beauty of mathematics, and in understanding the human aspects of the historical development of number."
—*Curriculum and Evaluation Standards for School Mathematics*

Many everyday situations cannot be adequately described without using both positive and negative numbers. Profit and loss, temperatures above and below 0°, elevations above and below sea level, and deposits and withdrawals are just a few examples. This chapter introduces negative numbers by extending your knowledge of whole numbers to the set of integers.

In Activities 1 and 4, ● represents a proton, and ○ represents an electron. Protons and electrons are subatomic particles. Protons have a positive electrical charge of one unit, and electrons have a negative electrical charge of one unit. Because they have opposite charges, when a proton and an electron are paired together, they neutralize each other; that is, the pair has an electrical charge of zero. You will use concrete representations for integers, like the charged particle model, and your understanding of the operations with whole numbers to develop algorithms for the integer operations.

Number theory is primarily concerned with the study of the properties of whole numbers. Number theory topics, such as multiples, factors, prime numbers, prime factorization, least common multiples, and greatest common divisors, are an integral part of the elementary mathematics curriculum. These concepts are used extensively when working with rational numbers. In this chapter, you will use concrete models to explore topics from number theory, and apply concepts in problem-solving situations.

Activity 1: Charged Particles

PURPOSE Develop a concrete model for integers and use it to explore absolute value.

MATERIALS Two different colored chips (or squares cut from tag board), 15 of each color

GROUPING Work individually or in groups of 2 or 3.

GETTING STARTED Use the dark chips to represent protons and the light chips to represent electrons. Construct two different models that represent each integer and sketch your models in the box.

Examples:

The set at the right shows one way to represent the number 2.

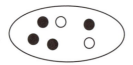

If the protons and electrons are paired, 2 protons are left over. The net electrical charge is 2.

The set at the right shows one way to represent the number –3.

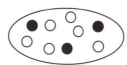

If the protons and electrons are paired, 3 electrons are left over. The net electrical charge is –3.

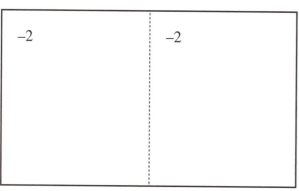

5	5

–1	–1

–2	–2

0	0

1. Look back at your models on the previous page. If you have not already done so, represent each integer using the fewest number of protons or electrons possible and sketch the model in the space provided.

    a. 5   b. −1

    c. −2   d. 0

2. What is the fewest number of particles needed to model each integer in Exercise 1?

    a. _____   b. _____   c. _____   d. _____

The answers to Exercise 2 are the absolute values of the integers in Exercise 1. The absolute value of an integer is the fewest number of protons or electrons that can be used to model it.

3. What is the absolute value of each of the following integers?

    a. −15 _____   b. 12 _____   c. 0 _____

4. Use your results from the preceding exercises to complete each statement.

    a. The absolute value of a positive integer is

    b. The absolute value of zero is

    c. The absolute value of a negative integer is

Activity 2: Coin Counters

PURPOSE   Investigate a concrete model for integers and use it to discover algorithms for integer addition.

MATERIALS   A paper cup, 10 pennies, and a game marker

GROUPING   Work in groups of 2 or 3.

GETTING STARTED
- At the beginning of the game, each player places a game marker on zero on a number line like the one below.
- Players alternate turns.
- On your turn, place six pennies in the cup, cover the opening with your hand, shake the cup thoroughly, and drop the coins onto the table. Each HEAD means move your marker to the right one unit; each TAIL means move it to the left one unit.
- The first player to go past 10 or –10 is the winner. If there is no winner after ten turns, the player closest to 10 or –10 wins.

Play the game twice. When you have finished, each player should answer the following questions.

1. Did you find a way to quickly determine where to place your marker after a coin toss? Explain.

2. If you were to represent the number of HEADS with an integer, would you use a positive or a negative integer?

3. Would you use a positive or a negative integer to represent the number of TAILS?

4. Did your marker ever end up an odd number of units away from where it was at the start of your turn? Explain.

5. At the end of a turn, did your marker ever end up in the same place where it started? Explain.

6. Use coins to construct two different representations for each integer. You may use more or less than six coins in a model.

   a. 4   b. –3   c. 0

You have seen how coins can be used to represent integers. Coins can also be used to model addition of integers. Think of the HEADS as a positive integer and the TAILS as a negative integer. For example, tossing 2 HEADS and 4 TAILS is the same as adding 2 and −4.

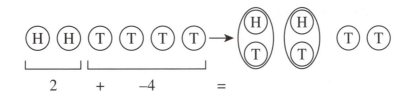

$$2 \quad + \quad -4 \quad =$$

1. a. Why do the paired coins cancel each other out?

    b. If you tossed this combination of coins, how would you move your marker?

    c. What integer is represented by the combination of coins?

    d. Complete the equation: $2 + -4 = $ _____.

2. Use coins to find the following sums. Make a sketch of your work. You may use more than 6 coins.

    a. $1 + -5$ b. $6 + -4$ c. $3 + -3$ d. $-5 + -2$

Use the coin model to answer the following questions.

3. a. Is the sum of two negative numbers positive or negative?

    b. How can you determine the sum of two negative numbers without using coins?

4. When is the sum of a positive and a negative number

    a. equal to 0?   b. positive?   c. negative?

5. How can you determine the sum of a positive and a negative number without using coins?

6. Use your rules from Exercises 3 and 5 to compute the following:

    a. $-17 + 25$ b. $13 + -7$ c. $-36 + -19$ d. $-11 + 11$

Activity 3: Addition Patterns

PURPOSE   Develop a rule for addition of integers by exploring patterns.

GROUPING   Work individually.

GETTING STARTED   Complete the sums you know, then look for patterns to fill in the missing entries in each list of problems. Use the completed lists to answer the questions.

4	+	3	=	7
4	+	2	=	6
4	+	1	=	____
4	+	0	=	____
4	+	–1	=	____
4	+	____	=	____
4	+	____	=	____
4	+	____	=	____
4	+	____	=	____

1. When is the sum of a positive number and a negative number

   a. positive?

   b. zero?

   c. negative?

2. Write a rule for adding a positive number and a negative number.

–4	+	2	=	–2
–4	+	1	=	–3
–4	+	0	=	____
–4	+	____	=	____
–4	+	____	=	____
–4	+	____	=	____
–4	+	____	=	____

Observe the pattern in the set of problems at the left and write a rule for the addition of two negative numbers.

EXTENSIONS   The following is a problem situation that illustrates addition of two positive integers. "Annie's allowance is $5 more than Michael's. Michael's allowance is $10 per week. What is Annie's allowance?" Write problem situations that illustrate the addition of (1) a positive and a negative integer and (2) two negative integers. Write two problems for each case.

Activity 4: Subtracting Integers

PURPOSE   Use the charged-particle model to develop a rule for subtracting integers.

MATERIALS   Two different colored chips (or squares), 15 of each color

GROUPING   Work individually or in groups of 2 or 3.

GETTING STARTED   Use the colored chips to represent protons and electrons. The following examples illustrate how to subtract integers using the charged-particle model.

Example 1:

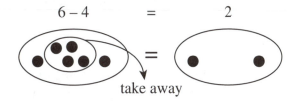

Example 2:

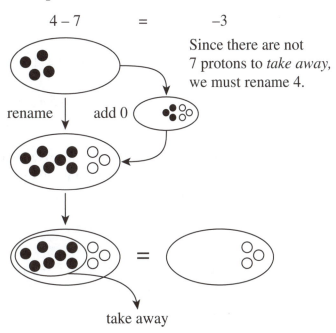

1. How could you rename –2 to compute the difference –2 – 5 using the charged-particle model?

2. Use the charged-particle model and colored chips to compute the following differences. Make a drawing to illustrate what you did in each problem.

   a. 5 – 9   b. 3 – –4   c. –2 – 5

   d. –6 – –5   e. –5 – 3   f. –4 – –8

3. Use the results from Exercise 2 to answer the following questions.

   a. When you subtract a positive integer from another integer, is the difference greater than or less than the original integer?

   b. When you subtract a negative integer from another integer, is the difference greater than or less than the original integer?

4. Determine the following:

   a. 5 + –9   b. 3 + 4   c. –2 + –5

   d. –6 + 5   e. –5 + –3   f. –4 + 8

5. How do the problems and answers in Exercise 4 a–f compare with the problems and answers in Exercise 2 a–f, respectively?

6. Study the comparisons in Exercise 5 to help write a rule for subtracting integers.

7. Use your rule from Exercise 6 to determine the following:

   a. –17 – –25   b. 13 – –7

   c. –36 – –19   d. –11 – 11

Activity 5: Subtraction Patterns

PURPOSE   Develop a rule for subtraction of integers by exploring patterns.

GROUPING   Work individually.

GETTING STARTED   In each of the following sets of problems, complete the differences you know, then look for patterns to fill in the missing entries.

1. 4 – 0 = 4
    4 – 1 = 3
    4 – 2 = ____
    4 – 3 = ____
    4 – ____ = ____
    4 – ____ = ____
    4 – ____ = ____
    ____ – ____ = ____

2. 3 – 4 = –1
    2 – 4 = –2
    1 – 4 = ____
    0 – 4 = ____
    ____ – 4 = ____
    ____ – ____ = ____
    ____ – ____ = ____
    ____ – ____ = ____

3. 4 – 3 = 1
    4 – 2 = 2
    4 – 1 = ____
    4 – ____ = ____
    4 – ____ = ____
    4 – ____ = ____
    ____ – ____ = ____
    ____ – ____ = ____

4. –4 – 3 = –7
    –4 – 2 = ____
    –4 – 1 = ____
    –4 – ____ = ____
    ____ – ____ = ____
    ____ – ____ = ____
    ____ – ____ = ____

5. Next to each of the subtraction problems above, write a related addition problem using numbers that have the same absolute value as those in the given problem.

Examples: $4 - 5 = 4 + -5$ $-4 - 1 = -4 + -1$

6. Write a rule for the subtraction of integers.

EXTENSIONS   Write problem situations that illustrate (1) subtraction of a negative integer from a positive integer and (2) subtraction of a negative integer from a negative integer. Write two problems for each case.

Activity 6: A Clown on a Tightrope

PURPOSE Develop rules for addition and subtraction of integers using a number line model.

MATERIALS A transparent copy of the large clown at the left.

GROUPING Work individually or in pairs.

GETTING STARTED The clown performs his tightrope act according to the following rules:

- He starts each act standing on the first number.
- For addition, the clown faces right.
- For subtraction, he faces left.
- A positive second number tells the clown to walk forward.
- A negative second number tells him to walk backward.

Example:

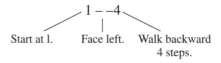

1. Where does the clown's act end in the Example? What does $1 - -4$ equal?

Use the tightrope below and the transparent copy of the clown to solve each problem. The arrow should always point to the clown's location on the rope.

2. $2 + 3 =$ _____

3. $3 + -5 =$ _____

4. $-4 + -2 =$ _____

5. $-7 + 9 =$ _____

6. $3 - 7 =$ _____

7. $6 - -2 =$ _____

8. $-5 - 3 =$ _____

9. $3 - -6 =$ _____

10. $-2 - -4 =$ _____

11. $-5 + 3 - -8 =$ _____

12. $2 - -7 + -12 =$ _____

13. $-9 - -11 - -7 =$ _____

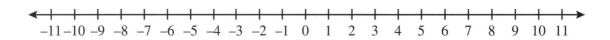

Use your results in Exercises 2–5 and the number line model to answer Exercises 14–16.

14. a. Is the sum of two negative integers positive or negative?

    b. How can you find the sum of two negative integers without using the number line model?

15. When is the sum of a positive and a negative integer

    a. positive?

    b. negative?

    c. equal to 0?

16. How can you determine the sum of a positive and a negative integer without using the number line model?

17. a. Write an addition problem in which the clown's starting point (3) and answer are the same as in Exercise 6.

    b. How are the numbers in the addition problem related to the numbers in the original problem?

18. Repeat Exercise 17 for each problem in Exercises 7–10.

19. Write a related addition problem that has the same answer as $-6 - -4$.

Activity 7: Multiplication and Division Patterns

PURPOSE   Develop rules for multiplication and division of integers by exploring patterns.

GROUPING   Work individually.

GETTING STARTED   In each of the following sets of problems, complete the products you know, then look for patterns to fill in the missing entries.

1.
4	×	3	=	12
4	×	2	=	___
4	×	1	=	___
4	×	___	=	___
4	×	___	=	___
___	×	___	=	___
___	×	___	=	___
___	×	___	=	___

2.
4	×	5	=	20
3	×	5	=	15
___	×	5	=	10
___	×	5	=	___
___	×	5	=	___
___	×	___	=	___
___	×	___	=	___
___	×	___	=	___

3.   Write a rule for multiplying a positive number and a negative number.

4.
−3	×	2	=	−6
−3	×	1	=	___
−3	×	0	=	___
−3	×	___	=	___
−3	×	___	=	___
___	×	___	=	___
___	×	___	=	___

5.
3	×	−6	=	___
___	×	−6	=	−12
___	×	−6	=	−6
___	×	___	=	___
___	×	___	=	___
___	×	___	=	___
___	×	___	=	___

6.   Write a rule for multiplying two negative numbers.

Recall that multiplication and division are *inverse operations*.

Example: $12 \div 4 = 3$, since $3 \times 4 = 12$. In general, $A \div B = C$ means that $C \times B = A$.

$-12 \div 4 = ?$ Think: $4 \times ? = -12$
                                        So $? = -3$.

Use the inverse relationship between multiplication and division to compute the quotients of several pairs of integers. Use the results to write a rule for division of integers.

Activity 8: A Square Experiment

PURPOSE Develop the concepts of prime, composite, and square numbers using a geometric model.

MATERIALS 30 squares per group

GROUPING Work individually or in groups of 3 or 4.

GETTING STARTED Use squares or graph paper to form all the rectangular arrays possible with each different number of squares. Record your results in Table 1.

Examples: Number of Squares   1   2   3

Rectangular Arrays

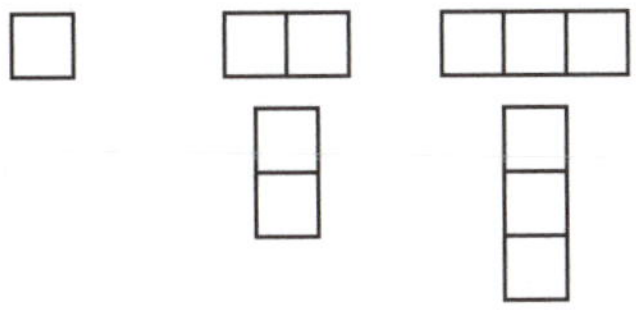

Note: 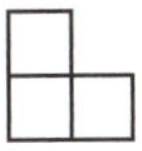 is not a rectangular array.

When an array is described by its dimensions, the figure 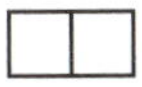 has an altitude of 1 unit and a base of 2 units and is labeled 1×2. The figure

is labeled 2×1.

TABLE 1

Number of Squares	Dimensions of the Rectangular Arrays	Total Number of Arrays
1		
2		
3	$1 \times 3, 3 \times 1$	2
4		
5		
6	$1 \times 6, 6 \times 1, 3 \times 2, 2 \times 3$	4
7		
8		
9		
10		
11		
12		

1. Use the results from Table 1 to complete Table 2.

TABLE 2

Number of Squares That Produced:			
A	**B**	**C**	**D**
Only One Array	**Only Two Arrays**	**More Than Two Arrays**	**An Odd Number of Arrays**

2. Suppose you have 24 squares.

   a. How many rectangular arrays can be made?

   b. In which column(s) in Table 2 would you place 24?

   c. What are the factors of 24?

3. a. What are the factors of 16?

   b. How many rectangular arrays can you make with 16 squares?

   c. In which column(s) of Table 2 would you place 16?

4. Look at the data in Table 1 and Table 2. How is the number of factors of a given number related to the number of rectangular arrays?

5. a. Why is it that the numbers in column D of Table 2 produce an odd number of arrays?

   b. What are the next three numbers that would be placed in column D?

6. What is the mathematical name for the numbers in

   a. column B?

   b. column C?

   c. column D?

7. Which numbers can be placed in two lists? Why?

8. Can any numbers be placed in three lists? If so, which ones?

9. Write each of the following composite numbers as a product of primes.

   a. 28   b. 42

   c. 150   d. 231

10. a. Can every composite number be written as a product of primes? Explain your reasoning.

    b. If two people write the same number as a product of primes,

       i. how would their factorizations be alike?

       ii. how might the factorizations be different?

Activity 9: A Sieve of Another Sort

PURPOSE   Investigate primes, composites, multiples, and prime factorizations using a "sieve."

MATERIALS   Orange, red, blue, green, and yellow colored pencils or crayons

GROUPING   Work individually.

GETTING STARTED   Eratosthenes, a Greek mathematician, invented the "sieve" method for finding primes over 2200 years ago. This activity explores a variation of Eratosthenes' sieve.

As you discovered in Activity 8, *one* is neither prime nor composite. To show this, mark an X through 1.

The first prime number is 2. Color the diamond in which 2 is located **orange**. Use **red** to color the upper-left corner of the **Key** and the upper-left corner of all squares containing multiples of 2. Any number with a corner colored will fall through the sieve.

What was the first multiple of 2 that fell through the sieve? _____

The next uncolored number is 3. Color the diamond surrounding the 3 **orange**. Use **blue** to color the upper-right corner of the **Key** and the upper-right corner of all squares containing multiples of 3.

What was the first multiple of 3 that fell through the sieve? _____

Repeat this process for 5 and 7. Color the diamonds surrounding the numbers **orange**. Use **green** to color the lower-right corners of the **Key** and of all squares containing multiples of 5. Use **yellow** to color the lower-left corners of the **Key** and of all the squares containing multiples of 7. Note the first multiples of 5 and of 7 that fall through the sieve.

Finally, use **orange** to color the diamond surrounding all the numbers in the grid that are in squares with no corners colored. These numbers are all primes.

1. How do you know that 2, 3, 5, and 7 are prime numbers?

2. How can you tell that 2, 3, 5, and 7 are prime numbers from the way the sieve is colored?

3. How can you identify composite numbers from the way the sieve is colored?

When you colored the multiples of 2, the number 4 was the first multiple of 2 that fell through the sieve.

When you colored the multiples of 3, the number 9 was the first multiple of 3 that fell through the sieve.

1. When you colored multiples, what was

   a. the first multiple of 5 that fell through the sieve?

   b. the first multiple of 7 that fell through the sieve?

After you colored the multiples of 7, the next uncolored number was 11.

2. If you could color multiples of 11, what would be the first number to fall through the sieve?

3. When you color multiples of a prime number, how is the first multiple of the prime that falls through the sieve related to the prime number?

4. If the grid went to 300, what is the largest prime whose multiples must be colored before you can be certain that all of the remaining uncolored numbers are prime?

5. What is the largest prime less than 1000? Explain how you obtained your answer.

The sieve can be used for more than finding primes.

1. List the numbers that are colored with the code for 2 and for 3.

2. The numbers in Exercise 1 are multiples of both 2 and 3. What numbers are they?

3. How could you use the color code to find

    a. the multiples of 14?

    b. the multiples of 30?

The sieve can also help you find the prime factorization of a number.

Example: From the sieve, you find that the $72 = 2 \times 3 \times 12 \leftarrow$ 12 is not prime
prime factors of 72 are 2 and 3.

$$2 \text{ is prime}$$
$$\downarrow$$

Again from the sieve, the prime $72 = 2 \times 3 \times (2 \times 3 \times 2)$
factors of 12 are 2 and 3. $= 2 \times 2 \times 2 \times (3 \times 3)$

Since 2 is a prime, you are done. $72 = 2^3 \times 3^2$

1. Find the prime factorization of

    a. 54   b. 84   c. 100

---

1. Pairs of prime numbers like 3 and 5 that differ by 2 are called *twin primes*. List all the twin primes less than 100.

2. What is the longest string of consecutive composite numbers on the grid?

3. Several of the numbers on the grid are divisible by three different primes. What is the smallest number that is divisible by four different primes?

Activity 10: The Factor Game

PURPOSE   Develop the idea of the prime factorization of a number using a game.

MATERIALS   Two different colored squares (page A-7), six paper clips, a calculator with fraction capability, Game Board, Factor List

GROUPING   Work in pairs or in teams of two students each.

GETTING STARTED   Use the game board and factor list on the next page. Play the Factor Game three times with your partner.

- To begin, one player places two paper clips on numbers in the factor list. The numbers may both be the same or they can be different. The player then multiplies the numbers and places a square of his or her color on the product on the game board.

- Players then alternate turns. On a turn, a player may form a new product in **one** of the following ways:

  A. Place a new paper clip on any number in the factor list.

  B. Move one of the paper clips already on the factor list to a different number.

  C. Take one paper clip off a number in the factor list.

- A turn ends when a player places a colored square on a product, or the product is already covered or it is not on the game board.

- The first player to cover three adjacent products in a row horizontally, vertically, or diagonally with squares of his or her color is the winner.

When you have finished playing the game, answer the following questions.

1. Are the numbers on the game board *prime* or *composite*? the numbers in the factor list?

2. a. List all of the factors of 60. Identify which factors are prime, which are composite, and which are neither prime nor composite.

   b. How could paper clips be placed on numbers in the factor list so that 60 could be covered on the game board?

   c. How many different ways can paper clips be placed on numbers in the factor list to result in a product of 60?

Factor Game

Game Board

4	12	25	90	36
108	75	120	6	10
15	80	8	9	45
54	30	100	18	50
20	24	27	40	60

Factor List

2   3   5

Every composite number can be written as the product of prime factors. Such a product is the **prime factorization** of the number.

A *factor tree* can be useful for finding the prime factorization of a number. An example of a factor tree for 45 is given at the right.

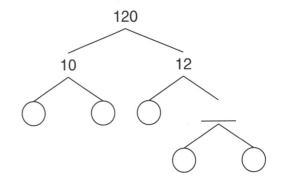

45

Step 1 → ⑤   9

Step 2 ⟶ ③   ③

1. Describe what happens in Step 1 of the factor tree for 45. Why is the 5 circled but not the 9?

2. Describe what happens in Step 2. Why do the branches stop at the circled numbers?

3. What is the prime factorization of 45?

---

1. Complete the factor tree for 120 at the right. Why are more branches necessary to make this tree than to make the one for 45?

120

10   12

2. What is the prime factorization of 120?

3. Sketch a factor tree of your own for 120 that starts with a different pair of factors.

4. Compare the circled numbers at the ends of the branches in the two factor trees for 120. What do you notice?

CALCULATORS CAN DO IT

The [Simp] key on calculators with fraction capability is used to express fractions in simplest form. It can also be used to find prime factorizations.

Example: Find the prime factorization of 18. (The key sequence shown is for the Texas Instruments Math Explorer. The key sequence may vary for other calculators.)

Key Sequence	Display	Record the Factors
ON/AC [1][8][/][1][8]	18/18	
[Simp] [=]	N/D → n/d 9/9	
[x ↔ y]	2	**2**
[x ↔ y] [Simp] [=]	N/D → n/d 3/3	
[x ↔ y]	3	**3**
[x ↔ y] [Simp] [=]	1/1	
[x ↔ y]	3	**3**
ON/AC [2][×][3][×][3][=]	18	**18 = 2 × 3 × 3**

1. Explain how the calculator procedure for finding the prime factorization and the factor tree method are related.

2. Find the prime factorization of each number.

   a. 72   b. 102   c. 210   d. 924

Activity 11: How Many Factors?

PURPOSE   Discover a rule for determining the number of factors of any number using patterns.

GROUPING   Work individually or in pairs.

Find the prime factorization (using exponential form) and the total number of factors for each number in the table.

Number	2	3	4	8	9
Prime Factorization					
Number of Factors					

Each of the numbers has only one prime factor. How is the total number of factors of each number related to the exponent in its prime factorization?

Use your answer to find a number that has 7 factors. _____ 11 factors. _____

Check your predictions by listing the factors of each number.

Find the prime factorization (using exponential form) and the total number of factors for each number in the table.

Number	6	24	60	72	100
Prime Factorization					
Number of Factors					

Each of the numbers in the table has more than one prime factor. How is the total number of factors of each number related to the exponents in its prime factorization?

Find the number of factors of 360 using the exponents in the prime factorization and by listing.

State a rule for determining the number of factors of any number from its prime factorization.

Use your answer to find a number that has 20 factors. Check your answer by listing the factors.

Activity 12: Interesting Numbers

PURPOSE   Review and apply the concepts of prime number, odd, even, perfect cube, and perfect square.

GROUPING   Work individually.

GETTING STARTED   The lyrics of an old song assert that "One is the loneliest number." This makes one interesting, but all numbers have some property that makes them interesting.

121

I'm a perfect square, since I'm equal to 11×11, but there is something else interesting about me. What is it?

1. Find three more perfect squares.

A *palindrome* is a number that is unchanged when its digits are reversed.

2. Find three more palindromes.

64

I'm equal to $4 \times 4 \times 4$. This means I'm a perfect cube, but I'm also a _____. This makes me doubly interesting!

1. Find three more perfect cubes.

2. What is the smallest multidigit number that is both a perfect cube and a palindrome?

13

I'm prime, since my only factors are 1 and 13, but I have another property that not all primes have. What is it? **HINT:** Reverse my digits.

1. How many palindromes between 100 and 200 are primes?

2. Find three prime numbers such that when their digits are reversed, the result is also a prime number.

Use the rating scale at the right to investigate some interesting three-digit numbers.

Example: 169

Perfect Square (169 = 13 × 13)	7 points
Sum of Digits Greater Than 14 (1 + 6 + 9 = 16)	4 points
Odd Number	2 points
Three Factors (1, 13, and 169)	3 points
Interest Rating	16 points

INTEREST RATING	
Prime Number	15 points
Perfect Cube	10 points
Perfect Square	7 points
Sum of Digits Greater Than 14	4 points
Even Number	3 points
Odd Number	2 points
Each Factor	1 point

1. Choose any three-digit number. Find its interest rating using the scale.

   Number _____
   Interest Rating _____

2. Can you find a three-digit number with an interest rating greater than 30 points? Explain.

3. What is the greatest possible interest rating for a three-digit prime number? Explain.

4. a. Explain why a number greater than one that is both a perfect cube and a perfect square would have an interest rating of at least 24 points.

   b. Are there any three-digit numbers that are both perfect cubes and perfect squares?

   c. If so, what are their interest ratings?

5. Try to find the three-digit number with the highest interest rating.

Activity 13: Tiling with Squares

PURPOSE Use a geometric model to explore the concept of the greatest common divisor (GCD) of two numbers and use the model to develop an algorithm for finding the GCD.

MATERIALS Half-centimeter graph paper (page A-48) and a calculator with fraction capability

GROUPING Work individually or in pairs.

GETTING STARTED Tiling a region means to completely cover it with non-overlapping shapes. The study of tilings can lead to some interesting questions. For example:

If m *and* n *are whole numbers, what is the length of a side of the largest square that can be used to tile an* m × n *rectangle?*

To answer this question, let's look at some rectangles. For each problem, draw the rectangle on graph paper and either draw or cut out squares to tile it.

1. In Parts a–d, if a 4 × 8 rectangle can be tiled with squares of the given size, make a sketch to show the tiling. If it can't, explain why not.

   a. 1 × 1   b. 2 × 2   c. 3 × 3   d. 4 × 4

2. Can squares of any other size be used to tile a 4 × 8 rectangle? Explain.

3. The results for a 4 × 8 rectangle are summarized in the table. Fill in the data for the other rectangles listed in the table.

Dimensions of Rectangle width × length	Prime Factorization		Common Factors	Dimensions of Squares That May Be Used	Length of Side of Largest Square
	width	length			
4 × 8	2 × 2	2 × 2 × 2	2 × 2	1 × 1, 2 × 2, 4 × 4	4
6 × 9					
18 × 30					
24 × 36					
8 × 15					

4. How does the length of a side of the largest square that can be used to tile a rectangle appear to be related to the common factors of the width and length of the rectangle?

1. a. List the divisors (factors) of 18.   b. List the divisors of 30.
    c. List the common divisors of 18 and 30.
    d. What is the *greatest common divisor* (GCD) of 18 and 30?

2. a. List the divisors of 24.   b. List the divisors of 36.
    c. List the common divisors of 24 and 36.
    d. What is the greatest common divisor (GCD) of 24 and 36?

3. How do the greatest common divisors found in Exercises 1 and 2 compare to the **Length of the Side of the Largest Square** that can be used to tile the corresponding rectangle in the table?

4. If m and n are whole numbers, what is the length of a side of the largest square that can be used to tile an $m \times n$ rectangle?

The $\boxed{\text{Simp}}$ key on calculators with fraction capability can be used to find greatest common divisors.

Example: Find the greatest common divisor of 54 and 72. (The key sequence shown is for the Texas Instruments Math Explorer. The key sequence may vary for other calculators.)

Key Sequence	Display	Record the Factors
ON/AC 5 4 / 7 2	54/72	
Simp =	27/36 N/D → n/d	
x ↔ y	2	**2**
x ↔ y Simp =	9/12 N/D → n/d	
x ↔ y	3	**3**
x ↔ y Simp =	3/4	
x ↔ y	3	**3**
ON/AC 2 × 3 × 3 =	18	GCD(54,72) = 2 × 3 × 3 GCD(54,72) = 18

1. Explain how the calculator procedure for finding the GCD and the method you found in Exercise 4 on the preceding page are related.

2. Find the GCD of each pair of numbers.
    a. 105 and 126   b. 38 and 95   c. 420 and 714   d. 924 and 770

Activity 14: Pool Factors

PURPOSE   Apply the concepts of greatest common divisor, least common multiple, and relatively prime numbers in a geometric problem situation.

MATERIALS   Graph paper and straightedge

GROUPING   Work individually or in groups of 3 or 4.

GETTING STARTED   On a piece of graph paper, draw several pool tables like the one shown below, but with different dimensions. Label the pockets *A*, *B*, *C*, and *D* in order, starting with the lower-left pocket as shown.

Place a ball on the dot in front of pocket *A*.

Shoot the ball as indicated by the arrows. The ball always travels on the diagonals of the grid and rebounds at an angle of 45 degrees when it hits a cushion.

Count the number of squares through which the ball travels.

Count the number of *hits,* that is, the number of times the ball hits a cushion, the initial hit at the dot, and the hit as the ball goes into a pocket.

In the table, enter the dimensions of each pool table, the number of squares through which the ball travels, and the number of hits. Analyze the data in the table and determine a rule that predicts the number of squares and the number of hits, given the dimensions of any pool table.

Height	Base	Number of Hits	Number of Squares
4	6		
5	7		
3	2		

Pool Table

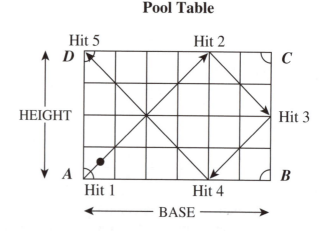

EXTENSIONS   Add a column headed **Final Pocket** to the table. In this column, for each pool table, record the letter of the pocket into which the ball finally fell. Use this data to find a rule that will predict which pocket the ball will fall into for any pool table.

Chapter 4 Summary

Like whole numbers, integers represent quantities. However, when you think about an integer, you usually think of it as representing not just a quantity, but also a direction. Thus you think of 5 and –5 as opposites, as a $5 profit and a $5 loss, or as 5 more than zero and 5 less than zero. This interpretation distinguishes integers from whole numbers and is reflected in the models used to represent integers. On a number line, 5 and –5 are both located 5 units from zero, but –5 is to the left of zero and 5 is to the right. When modeled as particles, in their simplest forms 5 and –5 are represented by the same number of particles, but the particles have opposite charges. These ideas were explored in Activities 1, 2, and 4.

The extension of the whole numbers to the integers required not only that you alter your concept of a number, but also that you modify your interpretations of the operations with numbers. Addition could still be thought of in terms of the union of sets. However, because the objects in the sets might be opposites, you found that in some cases you had to pair the opposites in order to find the sum.

Similarly, subtraction could still be thought of as *taking away*, but in some cases the objects in the sets representing the two numbers in the subtraction problem were opposites. In these cases, it was necessary to rename the number being subtracted before you could *take away*. The renaming was accomplished by adding zero. As a result, you discovered that subtraction can be interpreted as adding the opposite.

These changes in the meanings of addition and subtraction and the resulting algorithms were explored in Activities 2, 4, and 6. The results were verified in Activities 3 and 5 by examining patterns.

In Activity 7, patterns were analyzed to discover an algorithm for multiplying integers. The algorithm was extended to division by applying the inverse relationship between multiplication and division to find missing factors.

In Activities 8 and 9, you discovered that whole numbers can be classified by how many factors they have. *Prime numbers* have exactly two factors; *composite numbers* have more than two; *square numbers* have an odd number of factors; and *one*, which is in a class of its own, has exactly one factor.

Initially, this classification of the whole numbers may have seemed rather arbitrary. But in Activities 9 and 10, you discovered that every whole number greater than one can be expressed as a product of prime numbers, and that this product is unique except for the order of the factors. This means that, in a sense, the prime numbers are the building blocks from which all of the whole numbers greater than one are constructed. This result is so important that it is known as the *Fundamental Theorem of Arithmetic.*

In Activity 11, you discovered how the prime factorization of a number can be used to find the total number of factors of a given number. If you add 1 to each of the exponents in the prime factorization of the number, then the number of factors is the product of the sums. For example,

Prime factorization of $60 = 2^2 \cdot 3^1 \cdot 5^1$
Number of factors of $60 = (2 + 1) \cdot (1 + 1) \cdot (1 + 1)$
$$= 3 \cdot 2 \cdot 2$$
$$= 12$$

Activity 12 explored other interesting number classifications, such as *squares, cubes, palindromes,* and *emirps* (prime numbers, like 13, that are also prime when their digits are reversed), and Activities 13 and 14 were devoted to the study of factors and multiples of numbers. The *greatest common divisor* (GCD) and procedures for calculating it were developed using a geometric model and calculators. Least common multiples and greatest common divisors were applied in Activity 14 and will be used extensively in your study of the rational numbers.

Chapter 5
Rational Numbers
as Fractions

"Students should build their understanding of fractions as parts of a whole and as division. They will need to see and explore a variety of models of fractions, focusing primarily on familiar fractions such as halves, fourths, fifths, sixths, eighths, and tenths. By using an area model in which part of the region is shaded, students can see how fractions are related to a unit whole, compare fractional parts of a whole, and find equivalent fractions. They should develop strategies for ordering and comparing fractions, often using benchmarks such as $\frac{1}{2}$ or 1."

—Principles and Standards for School Mathematics

Other than whole-number computation, no topic in the elementary mathematics curriculum demands more time than the study of fractions. Yet, despite the years of study, most students enter high school with a poor concept of fractions and an even poorer understanding of operations with fractions. And when asked about their memories of fractions, adults will often reply, *"Yours is not to reason why, just invert and multiply."*

Rational numbers should be taught as the natural extension of the whole numbers. The fraction $\frac{3}{4}$ can be viewed as the solution to the problem of dividing 3 dollars among 4 people: $3 \div 4$. For students to understand this connection between whole numbers and fractions, teaching about fractions and their operations must be grounded in concrete models, as was the instruction with whole numbers. A well-developed concept of fractions and a feeling for their magnitude must be established before they can be compared and ordered meaningfully. This must precede computation.

A firm foundation of number sense involving fractions, and a deeper understanding of the algorithms for operations with fractions must be developed and precede formal work with fractions. This chapter provides activities which firmly establish the concept of fractions. All operations are explored through a variety of concrete models and reinforced by representing the models pictorially.

Activity 1: What Is a Fraction?

PURPOSE   Develop the concept of a fraction.
MATERIALS Pattern Blocks (pages A-7–A-11) and Cuisenaire Rods (page A-3)
GROUPING   Work individually or in pairs.

Use pattern blocks to solve the following problems.

1. The trapezoid is what fractional part of the hexagon?   _____

2. The blue rhombus is what fractional part of the hexagon?   _____

3. The triangle is what fractional part of the hexagon?   _____

4. The triangle is what fractional part of the blue rhombus?   _____

5. The triangle is what fractional part of the trapezoid?   _____

What fractional part of each figure is *shaded*? *unshaded*?

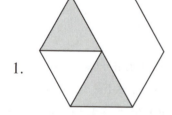

1.   shaded _____   unshaded _____

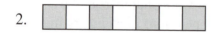

2.   shaded _____   unshaded _____

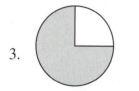

3.   shaded _____   unshaded _____

4.   shaded _____   unshaded _____

Use Cuisenaire rods to solve the following.

1. If the orange rod = 1, each rod is what fractional part of the orange rod?

   a. red _____   b. green _____

   c. yellow _____   d. purple _____

2. If the purple rod = 1, each rod is what fractional part of the purple rod?

   a. brown _____   b. orange _____

   c. dark green _____   d. black _____

---

1. If the red rod = $\frac{1}{2}$, which rod = 1? _____

2. If the red rod = $\frac{1}{3}$, which rod = 1? _____

3. If the white rod = $\frac{1}{5}$, which rod = 1? _____

4. If the white rod = $\frac{1}{4}$, which rod = $1\frac{3}{4}$? _____

5. If the red rod = $\frac{1}{2}$, which rod = $1\frac{1}{2}$? _____

6. If the red rod = $\frac{1}{3}$, which rod = $1\frac{2}{3}$? _____

Explain how you used the rods to arrive at your answers.

Use two red trapezoids and one blue rhombus to construct a shape similar to the one shown below.

1. Given that the shape = 1, what pattern block(s) would you use to represent each of the following fractions?

   a. $\frac{1}{4}$ _____   b. $\frac{1}{2}$ _____   c. $\frac{1}{8}$ _____

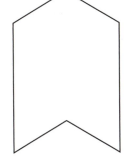

Fill in the same shape using one red trapezoid, two blue rhombuses, and one green triangle.

2. What fraction is represented by each of the following?

   a. a blue rhombus _____   b. a red trapezoid _____   c. a green triangle _____

Activity 2: Equivalent Fractions

PURPOSE   Develop the concept of a fraction using concrete models and a problem-solving approach.

MATERIALS   Cuisenaire Rods (page A-3), Pattern Blocks (pages A-7–A-11), and Fraction Strips (page A-31)

GROUPING   Work individually or in pairs.

For the following problems, use Cuisenaire rods to construct the trains.

1. Make all of the possible one-color trains the same length as a dark green rod and complete the following. If dark green = 1, then

   a. light green $= \dfrac{1}{2} = \dfrac{}{6}$

   b. red $= \dfrac{1}{} = \dfrac{}{6}$

   c. purple $= \rule{1cm}{0.4pt} = \rule{1cm}{0.4pt}$

   d. dark green $= \dfrac{}{6} = \rule{1cm}{0.4pt}$

   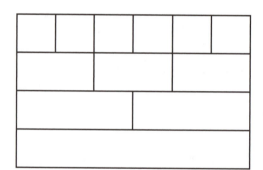

2. Make all of the possible one-color trains the same length as a brown rod and complete each of the following. If brown = 1, then

   a. purple $= \rule{1cm}{0.4pt} = \rule{1cm}{0.4pt} = \rule{1cm}{0.4pt}$

   b. dark green $= \rule{1cm}{0.4pt} = \rule{1cm}{0.4pt}$

   c. red $= \rule{1cm}{0.4pt} = \rule{1cm}{0.4pt}$

   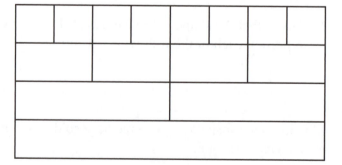

Use pattern blocks to construct a shape similar to the star and complete the following.

If the star shape = 1, then

1. trapezoid = $\dfrac{}{12}$ = $\dfrac{1}{}$

2. 2 blue rhombuses = $\dfrac{}{6}$ = $\dfrac{1}{}$ = $\dfrac{}{12}$

3. hexagon = $\dfrac{6}{}$ = $\dfrac{}{6}$ = $\dfrac{1}{}$

Use the fraction strips to find strips that can be folded into parts so that the resulting strip is equal in length to the fraction given in each problem. Folds may be made **only** on the lines on the strips.

Write the name of the equivalent fractions in the space provided.

1. $\dfrac{1}{2}$ = _____ = _____ = _____ = _____

2. $\dfrac{2}{3}$ = _____ = _____ = _____

3. $\dfrac{3}{4}$ = _____ = _____

EXTENSIONS

1. Given a set of fractions equivalent to the fraction $\dfrac{a}{b}$, where $\dfrac{a}{b}$ is in lowest terms, what is the relationship among the set of numerators? the set of denominators?

2. Given a set of equivalent fractions, $\dfrac{a}{b}, \dfrac{c}{d}, \dfrac{e}{f}, \ldots$, where $\dfrac{a}{b}$ is in lowest terms, what is the relationship between the numerator and denominator of $\dfrac{a}{b}$ and the numerator and denominator of any equivalent fraction?

Activity 3: How Big Is It?

PURPOSE   Develop the ability to estimate the magnitude of a fraction.

MATERIALS   A deck of fraction cards (pages A-32 and A-33) and a copy of the Fraction Sorting Board (page A-34)

GROUPING   Work individually or in pairs.

GETTING STARTED   Use these rules to complete the following problems.

A fraction is close to   **1**   if the numerator and denominator are approximately the same size.

$\dfrac{1}{2}$ if the denominator is about twice as large as the numerator.

0   if the numerator is very small compared to the denominator.

THE FRACTION SORTING GAME

This is a game for two players. Cut out the fraction cards and shuffle them. Students take turns placing a card in the appropriate column on the fraction sorting board and justifying each placement to the other player. Reshuffle the deck and play again.

1. Complete the following fractions so that they are close to, but less than, $\dfrac{1}{2}$.

   a. $\dfrac{\rule{1cm}{0.4pt}}{100}$ b. $\dfrac{\rule{1cm}{0.4pt}}{25}$ c. $\dfrac{\rule{1cm}{0.4pt}}{9}$ d. $\dfrac{\rule{1cm}{0.4pt}}{14}$

   e. $\dfrac{7}{\rule{1cm}{0.4pt}}$ f. $\dfrac{3}{\rule{1cm}{0.4pt}}$ g. $\dfrac{11}{\rule{1cm}{0.4pt}}$ h. $\dfrac{8}{\rule{1cm}{0.4pt}}$

2. Complete the following fractions so that they are close to, but less than, 1.

   a. $\dfrac{\rule{1cm}{0.4pt}}{27}$ b. $\dfrac{\rule{1cm}{0.4pt}}{12}$ c. $\dfrac{\rule{1cm}{0.4pt}}{75}$ d. $\dfrac{\rule{1cm}{0.4pt}}{8}$

   e. $\dfrac{9}{\rule{1cm}{0.4pt}}$ f. $\dfrac{3}{\rule{1cm}{0.4pt}}$ g. $\dfrac{11}{\rule{1cm}{0.4pt}}$ h. $\dfrac{95}{\rule{1cm}{0.4pt}}$

EXTENSIONS   1. Given the fraction $\dfrac{13}{\square}$, what numbers would be acceptable in place of $\square$ so that the resulting fraction is *close to but less than 1*? Explain your answer.

2. Given the fraction $\dfrac{\square}{23}$, what numbers would be acceptable in place of $\square$ so that the resulting fraction is *close to $\dfrac{1}{2}$*? Explain your answer.

Activity 4: Fraction War

PURPOSE   Reinforce estimation and comparison of fractions in a game format.

MATERIALS   A deck of playing cards (remove the face cards), the Fraction Arrays (page A-35), and the Fractions Game Board (page A-36)

GROUPING   Work in pairs.

GETTING STARTED   Rules for Fraction War.

Example:

FRACTIONS GAME BOARD

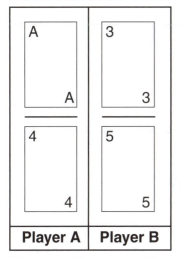

Player A wins a round in Game a: *the smaller fraction* is the winner.

1. Shuffle the cards and deal them evenly, face down to each player. Players choose a goal for a game from those listed below.

    a. Form a proper fraction by placing the card with the smaller number in the numerator. Player with the smaller fraction is the winner.

    b. Form a proper fraction by placing the card with the smaller number in the numerator. Player with the larger fraction is the winner.

    c. Form a proper fraction by placing the card with the smaller number in the numerator. Player with the fraction whose value is closest to $\frac{1}{2}$ is the winner.

    d. Form a fraction by placing the card with the larger number in the numerator. Player with the larger fraction is the winner.

    e. Place the first card in the numerator, the second in the denominator. Player with the fraction whose value is closest to 2 is the winner.

    f. Each player decides where to place each card. Player with the fraction whose value is closest to 1 is the winner.

2. Each player turns up two cards from his or her pile and follows the directions for the chosen game to form a fraction on the Fractions Game Board. The ace represents 1.

3. The winner of each round collects the four cards and places them face up at the bottom of his or her pile of cards. If the fractions formed are equivalent, each player turns over two additional cards and forms a new fraction. The winner of the round gets all eight cards.

4. When the players have played all the face-down cards, the player with the most face-up cards is the winner of the game. Reshuffle the cards and choose a different game.

Activity 5: What Comes First?

PURPOSE   Develop an understanding of comparing and ordering fractions.

MATERIALS   Cuisenaire Rods (page A-3), Pattern Blocks (pages A-7–A-11), Fraction Arrays (page A-35)

GROUPING   Work individually.

Use Cuisenaire rods to build all possible one-color trains that are the same length as a brown rod, and complete the following.

1. Which is larger, $\dfrac{5}{8}$ or $\dfrac{1}{2}$? _____   Complete the inequality: _____ > _____

2. Which is smaller, $\dfrac{3}{8}$ or $\dfrac{1}{4}$? _____   Complete the inequality: _____ < _____

3. Which is larger, $\dfrac{7}{8}$ or $\dfrac{3}{4}$? _____   Complete the inequality: _____ > _____

Use pattern blocks to construct a star shape (see Activity 2) and complete the following.

1. Which is larger, $\dfrac{5}{12}$ or $\dfrac{1}{2}$? _____   Complete the inequality: _____ < _____

2. Which is smaller, $\dfrac{2}{3}$ or $\dfrac{7}{12}$? _____   Complete the inequality: _____ > _____

3. Which is larger, $\dfrac{5}{6}$ or $\dfrac{3}{4}$? _____   Complete the inequality: _____ < _____

Use the fraction arrays to order the following fractions.

1. $\dfrac{1}{2}, \dfrac{3}{5}, \dfrac{4}{7}$ _____ > _____ > _____

2. $\dfrac{2}{3}, \dfrac{3}{4}, \dfrac{7}{8}$ _____ > _____ > _____

3. $\dfrac{5}{12}, \dfrac{2}{5}, \dfrac{3}{7}$ _____ < _____ < _____

4. $\dfrac{5}{6}, \dfrac{11}{12}, \dfrac{4}{5}$ _____ < _____ < _____

Activity 6: Square Fractions

PURPOSE   Reinforce the concept of a fraction and the meaning of equivalent fractions, and illustrate operations with fractions using geometric models.

MATERIALS   A sheet of paper 20 cm square (colored construction paper works well), and scissors (optional)

GROUPING   Work individually or use as a class activity directed by the instructor.

GETTING STARTED   Work through each section of the activity in order, following the folding and cutting directions carefully. If scissors are not used, fold and crease the paper sharply so that it will tear cleanly.

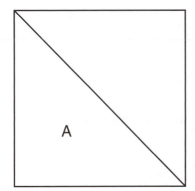

1. Fold the square as shown and cut or tear along the fold to divide the square into two congruent parts.

    Each polygon is what fraction of the original square? _____

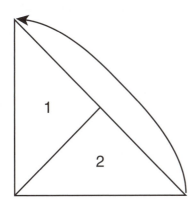

Pick one of the two triangles and fold it as shown. Cut or tear along the fold and label the two triangles 1 and 2.

2. Triangle 1 is what fractional part of

    a. triangle A? _____

    b. the original square? _____

    Explain your answers.

In the remaining large triangle, fold the vertex of the right angle to the midpoint of the longest side. Cut along the fold and label the polygons B and 3 as shown.

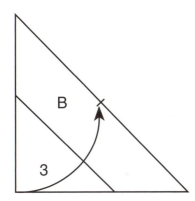

3. Triangle 3 is what fractional part of

    a. triangle 1? _____

    b. triangle A? _____

    c. trapezoid B? _____

Fold one of the endpoints of the longest side of trapezoid B to the midpoint of that side as shown. Cut along the fold and label the polygons C and 4.

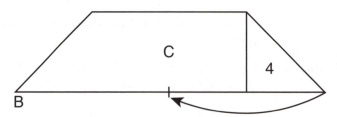

4. Triangle 4 is what fractional part of

    a. triangle 3? _____

    b. triangle A? _____

    c. trapezoid C? _____

Fold trapezoid C as shown. Cut along the fold and label the polygons D and 5.

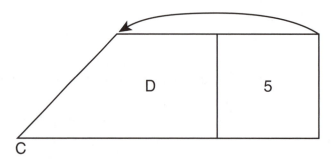

5. Square 5 is what fractional part of

    a. trapezoid C? _____

    b. trapezoid D? _____

    c. triangle 3? _____

    d. the original square? _____

6. Trapezoid C is what fractional part of

    a. triangle A? _____   b. square 5? _____

7. Trapezoid D is what fractional part of

    a. triangle 1? _____   b. trapezoid C? _____

Fold trapezoid D as shown. Cut along the fold and label the two polygons 6 and 7.

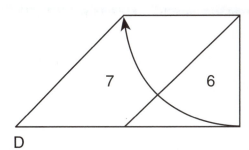

D

8. Triangle 6 is what fractional part of

    a. trapezoid D? _____

    b. triangle 3? _____

9. Parallelogram 7 is what fractional part of

    a. square 5? _____   b. trapezoid B? _____

    c. triangle 1? _____   d. original square? _____

10. Explain how triangle 1, triangle A, and the original square can be used to illustrate that $\dfrac{2}{4}$ is equivalent to $\dfrac{1}{2}$.

11. Explain how these figures can be used to illustrate that $\dfrac{1}{2}$ of $\dfrac{1}{2}$ is $\dfrac{1}{4}$, which is the multiplication problem $\dfrac{1}{2} \times \dfrac{1}{2} = \dfrac{1}{4}$.

12. If the original square is one unit, explain how trapezoid D and triangle 3 can be used to model the addition problem

    $\dfrac{3}{16} + \dfrac{1}{8} = \dfrac{5}{16}$.

13. If trapezoid B is one unit, explain how triangle 4 and trapezoid D can be used to model the division problem

    $\dfrac{1}{2} \div \dfrac{1}{6} = 3$.

Activity 7: Adding and Subtracting Fractions

PURPOSE   Develop algorithms for adding and subtracting fractions using concrete models.

MATERIALS   Pattern Blocks (pages A-7–A-11)and Fraction Strips (page A-31)

GROUPING   Work individually.

If the yellow hexagon = 1, then the red trapezoid = $\frac{1}{2}$, the blue rhombus = $\frac{1}{3}$, and the green triangle = $\frac{1}{6}$. Use pattern blocks to solve the following:

1. 1 red + 3 green   =   1 red + 1 red   =   3 green + 3 green   =

$\frac{1}{2}$ + $\frac{3}{6}$ = $\frac{1}{2}$ + $\frac{1}{2}$ = $\frac{3}{6}$ + $\frac{3}{6}$ = _____

2. 1 red + 1 blue   = ___ green + ___ green   =   _____

$\frac{1}{2}$ + $\frac{1}{3}$ = _____ + _____   = _____

3. 1 blue + 1 green = _____ + _____   =   _____

$\frac{1}{3}$ + $\frac{1}{6}$ = _____ + _____   = _____   = _____

4. 1 red – 1 blue   = _____ – _____   =   _____

$\frac{1}{2}$ – $\frac{1}{3}$ = _____ – _____   = _____

5. 1 red – 1 green   = _____ – _____   =   _____

$\frac{1}{2}$ – $\frac{1}{6}$ = _____ – _____   = _____   = _____

Use pattern blocks to solve the following problems. Write your answers in simplest form, that is, the number represented by the least number of blocks of the same color.

Let  one whole   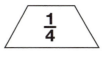 $\frac{1}{2}$ $\frac{1}{4}$ $\frac{1}{6}$ $\frac{1}{12}$

1. $\dfrac{1}{2} + \dfrac{3}{12} =$

2. $\dfrac{3}{4} + \dfrac{1}{2} + \dfrac{1}{6} =$

3. $\dfrac{3}{4} - \dfrac{2}{3} =$

4. $\dfrac{5}{6} - \dfrac{3}{4} =$

5. $\dfrac{2}{3} + \dfrac{1}{2} =$

6. $\dfrac{3}{4} + \dfrac{2}{3} + \dfrac{1}{6} =$

7. $1\dfrac{1}{2} + \dfrac{2}{3} =$

8. $1\dfrac{5}{12} - \dfrac{5}{6} =$

When adding fractions using fraction strips, you must fold the strips to show only the fractions that are needed. Strips are placed as shown in the following figures. A longer strip must then be found that has fold lines in common with the two fractions.

Example for Addition: $\dfrac{1}{3} + \dfrac{1}{4} =$

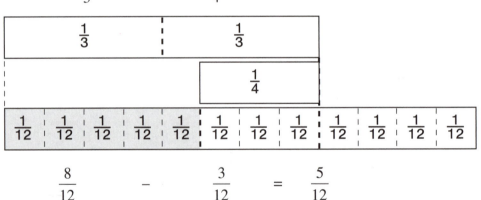

$$\frac{1}{3} \qquad + \qquad \frac{1}{4}$$

$$\frac{4}{12} \qquad + \qquad \frac{3}{12} \qquad = \qquad \frac{7}{12}$$

Example for Subtraction: $\dfrac{2}{3} - \dfrac{1}{4} =$

$$\frac{2}{3} \qquad - \qquad \frac{1}{4} \qquad =$$

$$\frac{8}{12} \qquad - \qquad \frac{3}{12} \qquad = \qquad \frac{5}{12}$$

Use your fraction strips to solve the following problems.

1. $\dfrac{1}{2} + \dfrac{3}{5} =$ 2. $\dfrac{7}{8} - \dfrac{1}{4} =$

3. $\dfrac{7}{10} - \dfrac{2}{5} =$ 4. $\dfrac{5}{12} + \dfrac{1}{3} =$

5. $\dfrac{2}{3} + \dfrac{3}{4} =$ 6. $\dfrac{3}{4} - \dfrac{1}{6} =$

Activity 8: Multiplying Fractions

PURPOSE	Develop an algorithm for multiplying fractions using Pattern Blocks and paper folding.
MATERIALS	Pattern Blocks (pages A-7–A-11) and paper for folding
GROUPING	Work individually.

Example: $\frac{2}{3}$ of $\frac{1}{4}$ means two of three equal parts of $\frac{1}{4}$.

 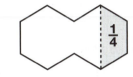 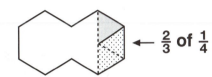

$$\frac{2}{3} \times \frac{1}{4} = \frac{2}{12} = \frac{1}{6}$$

Place pattern blocks on Figure A to solve the following. Record your solution both pictorially and numerically.

Figure A

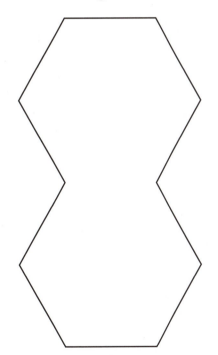

1. $\frac{1}{2} \times \frac{1}{3} =$

2. $\frac{3}{4} \times \frac{1}{3} =$

3. $\frac{1}{4} \times \frac{1}{3} =$

4. $\frac{3}{4} \times \frac{2}{3} =$

5. $\frac{5}{6} \times \frac{1}{2} =$

Example: $\frac{1}{2}$ of $\frac{2}{3}$ means one of the two equal parts of two thirds.

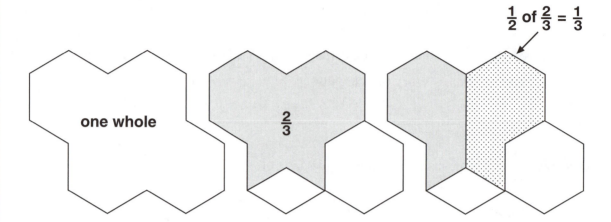

Use four hexagons to construct a figure similar to the one shown above, and solve the following. Record each step of your solutions both pictorially and numerically.

1. $\dfrac{3}{4} \times \dfrac{1}{6} =$

2. $\dfrac{3}{8} \times \dfrac{2}{3} =$

3. $\dfrac{7}{12} \times \dfrac{1}{2} =$

4. $\dfrac{5}{8} \times \dfrac{1}{3} =$

$\frac{1}{3}$ of 1 means one of the three equal parts of 1. Divide a piece of paper into thirds with vertical folds.

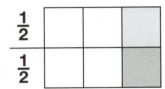

$\frac{1}{2}$ of $\frac{1}{3}$ means one of the two equal parts of $\frac{1}{3}$. Now, divide the thirds into halves with a horizontal fold.

$$\frac{1}{2} \text{ of } \frac{1}{3} = \frac{1}{6}$$

$\frac{2}{3}$ of $\frac{1}{2}$ means two of the three equal parts of $\frac{1}{2}$.

$$\frac{2}{3} \text{ of } \frac{1}{2} = \frac{2}{6} = \frac{1}{3}$$

Fold sheets of paper to solve the following. Record each step of your solution pictorially and numerically.

1. $\frac{1}{3} \times \frac{3}{4} =$

2. $\frac{1}{4} \times \frac{2}{3} =$

3. $\frac{2}{3} \times \frac{2}{3} =$

4. $\frac{3}{8} \times \frac{1}{3} =$

EXTENSIONS   From what you have observed in this activity, write a rule for multiplication of fractions.

Activity 9: Dividing Fractions

PURPOSE   Develop the understanding of division of fractions using pattern blocks.

MATERIALS   Pattern blocks (pages A-7–A-11)

GROUPING   Work individually.

GETTING STARTED   Recall the use of the multiplication and division frame for the division of whole numbers.

Example: $3\overline{)6}$ can mean how many groups of 3 are there in 6? $3\overline{\smash{\big|}}\begin{smallmatrix}2\\ \boxplus\end{smallmatrix}$

In the following example $\bigcirc$ represents 1.

Example: $1 \div \dfrac{1}{2}$ means: How many groups of $\dfrac{1}{2}$ are there in 1?

$1 \div \dfrac{1}{2} = 2$

Complete each sentence and use your pattern blocks to solve the following problems.

1. $\dfrac{1}{3} \div \dfrac{1}{6}$ means _____

   $\dfrac{1}{3} \div \dfrac{1}{6} =$

2. $\dfrac{1}{2} \div \dfrac{1}{4}$ means _____

   $\dfrac{1}{2} \div \dfrac{1}{4} =$

3. $\dfrac{5}{6} \div \dfrac{5}{12}$ means _____

   $\dfrac{5}{6} \div \dfrac{5}{12} =$

4. $\dfrac{3}{4} \div \dfrac{1}{4}$ means _____

   $\dfrac{3}{4} \div \dfrac{1}{4} =$

5. $\dfrac{3}{2} \div \dfrac{3}{4}$ means _____

   $\dfrac{3}{2} \div \dfrac{3}{4} =$

● To model the problem $\frac{1}{2} \div \frac{1}{3}$, let

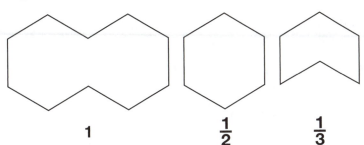

1                      $\frac{1}{2}$ $\frac{1}{3}$

How many sets of $\frac{1}{3}$ (two blue rhombuses) are there in $\frac{1}{2}$ (one hexagon)?

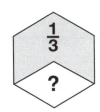

There is one group of $\frac{1}{3}$ (two blue rhombuses), plus a remainder.

● The remainder is equal to one blue rhombus, which is $\frac{1}{2}$ of $\frac{1}{3}$.

Therefore $\frac{1}{2} \div \frac{1}{3}$ = one set of two blue rhombuses + one half set of two blue rhombuses

=   1   + $\frac{1}{2}$

= $1\frac{1}{2}$

Use your pattern blocks to solve the following:

1. $\frac{5}{6} \div \frac{1}{3} =$ 2. $\frac{3}{4} \div \frac{1}{2} =$

3. $\frac{2}{3} \div \frac{1}{2} =$ 4. $1\frac{1}{3} \div \frac{1}{2} =$

● **EXTENSIONS**   Describe how you would use fraction strips to solve the previous problems. Draw at least one illustration of your method.

Activity 10: People Proportions

PURPOSE Reinforce the concept of ratio and apply ratios in a real-world setting that makes connections to science.

MATERIALS A calculator and two 150-cm measuring tapes for each group of students

GROUPING Work in groups of 4 or 5.

GETTING STARTED To gain an understanding of the relationships among body measurements, measure your hand span—the largest spread between the tip of the thumb and the tip of the middle finger. Wrap your hand around your wrist to compare the hand span with the circumference of the wrist; they should be about equal. Place both hands around your neck. The circumference of your neck is about twice your hand span, or twice the circumference of your wrist.

Measure the body parts listed below for each student in your group. All measurements should be in centimeters and rounded to the nearest 0.5 cm. Follow the directions below when making the measurements.

A. Height and navel to floor: Measure without shoes.

B. Wing span: Measure fingertip to fingertip with arms outstretched.

C. Tibia: Rest the foot on the floor and measure from the ankle bone to the top of the tibia on the outside of the kneecap.

D. Radius: Rest the elbow on the desk with the hand up and measure from the tip of the elbow to the wrist bone.

Enter the name of the person and each measurement for that person in Table 1.

TABLE 1

Name of Person	Height	Navel to Floor	Wing Span	Wrist	Radius	Neck	Tibia

Use the measurements from Table 1 to write the fractional form of
the ratios for each person in the columns labeled F in Table 2. Do not
convert the fractions to decimals.

TABLE 2

Name of Person	Wing Span / Height		Height / Navel to Floor		Tibia / Height		Radius / Height		Wrist / Neck	
	F	D	F	D	F	D	F	D	F	D

1. Examine the ratios for the people in your group. Are any of the
   ratios nearly the same for all people? _____
   If so, which ones?

2. Now convert all the fractions to decimals, rounded to the nearest
   thousandth. Compare the decimal equivalents for each of the
   ratios. Which ratios are approximately the same for all people?

3. Find the average (mean) of the decimal ratios for tibia to height
   and radius to height.

   a. The length of the tibia is approximately what fraction of
      the height? _____

   b. The length of the radius is approximately what fraction of
      the height? _____

4. Find the average of the ratios for height to navel-to-floor as a
   decimal. _____

   For navel-to-floor to height. _____

5. The average ratio for height to navel-to-floor for all humans is approximately 1.618. Compare this decimal with the average from your group in Exercise 4.

6. Write the reciprocal for the decimal 1.618 rounded to the nearest thousandth. ____

   What did you find?

   Can you find any other decimal with this property?

---

1. Paleontologists may find only a few bones of an ancient person, yet be able to reconstruct an entire skeleton. A forensic medical examiner may provide police with the height, weight, age, sex, and other features of a body from as little evidence as a skull or two or three bones. How can the skeleton of a human be reconstructed with so little evidence?

2. You found that several ratios for body measurements for all the people in your group were approximately the same. Would you expect the same results for people of any age, size, or shape?

3. Make the same measurements for several people whose age or size differs from the people in your group. Determine if the ratios are about the same as those of your group.

4. Research the Golden Ratio. Compare the Golden Ratio to your answer in Exercise 4. Find other body ratios that are close to the Golden Ratio. Describe at least one application of the Golden Ratio in art, music, nature, architecture, psychology, phyllotaxis, or sea shells.

WHY NOT?

Use a proportion to solve each of the following:

1. Josh Jumper can long jump 5 m with a 15-m run. How far could he jump with a 75-m run? a 150-m run?

2. Wally Walker says he can walk 6 km in an hour. How long would it take him to walk 60 km? 150 km?

3. Tall Tanya was 145 cm tall when she was 12 years old. When she was 15, she was 165 cm tall. How tall will she be when she is 24? 30?

For each problem, explain why solving the proportion produces results that do not fit the real-life situation.

Activity 11: A Call to the Border

PURPOSE   Apply ratio and proportion, measurement, scale interpretation, estimation, and communication in a problem-solving context.

MATERIALS   Almanac or atlas, state map, ruler, string, scissors, and calculator

GROUPING   Work in groups of four.

GETTING STARTED

The Daily Tabloid

Governor Cuts State Spending
Eliminates National Guard

The governor defends this action by claiming that if he ordered every man, woman, and child to go to the border of the state, all residents could stand side-by-side around the state to defend the border.

Do you think the governor's claim is true?

If the residents of your state lined up along the state's border, do you think people standing next to each other would be able to:

- See each other?
- Carry on a conversation?
- Touch fingertip-to-fingertip?

- Hold hands?
- Stand shoulder-to-shoulder?
- Stand belly-to-back?

1. List the information you will need to verify the governor's claim and your prediction of how close people will be able to stand.

2. Explain how the perimeter of your state is affected by irregularities such as coastline, islands, river boundaries, mountains, etc.

3. Record the following:

    State _____   Population _____   Perimeter of state _____

4. Is the governor's claim correct? Why or why not? Explain your reasoning.

EXTENSIONS   In which states might this solution to state spending cuts NOT work? Which states pose special problems in implementing your method for solving the problem? Name some states where residents might be standing closer together or farther apart than in your state.

Chapter 5 Summary

The activities in this chapter emphasized the development of a conceptual understanding of rational numbers (fractions) and their operations. Concrete materials and structured lessons illustrated how the operations with fractions are an extension of the operations with whole numbers.

A curriculum developmentally appropriate for students is one of the central themes of the *Principles and Standards for School Mathematics*, NCTM 2000. This implies that lessons progress from the concrete to the representational (pictorial) to the abstract. The activities in this chapter illustrated one aspect of such a curriculum through careful development of the operations with fractions.

Activity 1 developed the concept of a fraction using a variety of models. In different problems, you (a) determined the fractional part of a whole, (b) compared two areas or the length of two strips to determine a fraction, and (c) determined the whole unit given the fractional part.

Activity 2 introduced equivalent fractions using several models. The concept of equivalence is critical to understanding ordering fractions, and adding and subtracting fractions. Activities 3–5 may be the most important ones in the chapter. They addressed another central theme of the *Principles and Standards for School Mathematics:* developing number sense. Only when the concept of a fraction is understood, can one develop a sense of their size. Knowing terms like *about the same size, half as much,* and *very small as compared to* is all that is necessary to estimate the size of any fraction.

Activity 6 asked you to use geometric representations of fractions which you constructed by folding and cutting a square. The pieces of the square helped you to explore the relationship among various pieces that made up the whole, and reinforced the understanding of addition, subtraction, multiplication and division of fractions.

Activity 7 used three models to explore addition and subtraction of fractions. Each one used the concept of equivalence as developed in the previous activities. The importance of common denominators was connected to dividing the whole into equivalent parts that could then be added or subtracted. Activity 8 illustrated the language relationship between the word *of* and multiplication of rationals through a geometric model for fractions and, Activity 9 modeled division of fractions in the same way that division of whole numbers was shown. That is, how many groups of one factor (the divisor) are there in the product (the dividend). By modeling division this way, you can come to understand the familiar rule, "invert and multiply."

Activity 10 applied the concepts of ratio and proportion to human anatomy. You learned that several ratios comparing selected body parts are approximately the same regardless of a wide variation in people's height or weight. You discovered the Golden Ratio during this activity. It can be found in hundreds of comparisons in the human body, in many other applications in nature, as well as in art, architecture, music, and psychology.

Activity 11 involved a rich problem solving situation in which proportional reasoning was applied to scale interpretation on maps. Using reference materials to determine the state population, estimating the space a person can occupy, and finding the best estimate for the perimeter of a state were all required to solve the Border problem.

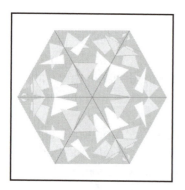

Chapter 6
Decimals, Precents, and Real Numbers

"The foundation of students' work with decimal numbers must be an understanding of whole numbers and place value. Students should learn to think of decimal numbers as a natural extension of the base-ten place-value system to represent quantities less than 1. They should also understand decimals as fractions whose denominators are powers of ten. The absence of a solid conceptual foundation can greatly hinder students. Students also need to interpret decimal numbers as they appear on calculator screens, where they may be truncated or rounded.

Percents, which can be thought about in ways that combine aspects of both fractions and decimals, offer students another useful form of rational numbers. Percents are particularly useful when comparing fractional parts of sets or numbers of unequal size, and they are also frequently encountered in problem-solving situations that arise in everyday life. As with fractions and decimals, conceptual difficulties need to be carefully addressed in instruction."

—*Principles and Standards for School Mathematics*

Instruction related to decimals and percents and computation with decimals will have its greatest impact when it is based on the same models and understanding as fractions and whole numbers.

The activities in this chapter extend and reinforce the models which have been used previously with whole numbers and fractions. Decimal numbers will be modeled with base-ten blocks in a variety of ways.

Example: *If a rod = 1, then the small cube = 0.1.*
          If a flat = 1, then a rod = 0.1, and the small cube = 0.01

These models reinforce the concept of a decimal number being part of a whole.

Number and operation sense and estimation skills with whole numbers developed in Chapters 2 and 3 also will be reviewed and applied in the development of algorithms for multiplication of decimals.

120

The activities in this chapter continue the extension of whole number concepts to percent by applying models that have been used previously with whole numbers, fractions, and decimals. The word percent means part of 100. The idea of part of a whole relates to the concept of a fraction. When the whole is 100, there is also a direct connection to decimals.

Concepts of ratio and proportion will be studied in real world applications that make connections to science, nature, and geography.

Activity 1: What's My Name?

PURPOSE	Develop the relationship between fractions and decimals.
MATERIALS	Base Ten Blocks (pages A-13 and A-15)
GROUPING	Work individually.
GETTING STARTED	Let 1 rod = 1, and 1 cube = 0.1.

Write the fraction and the decimal for the shaded part of each figure. If necessary, use the cubes to determine the decimal part.

Fraction   Decimal

1. _____ _____

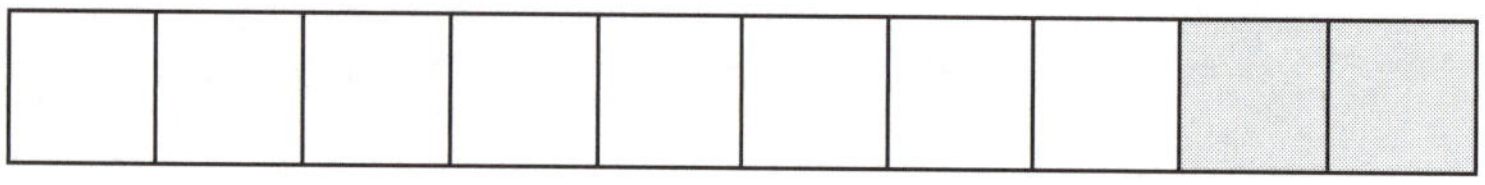

2. _____ _____

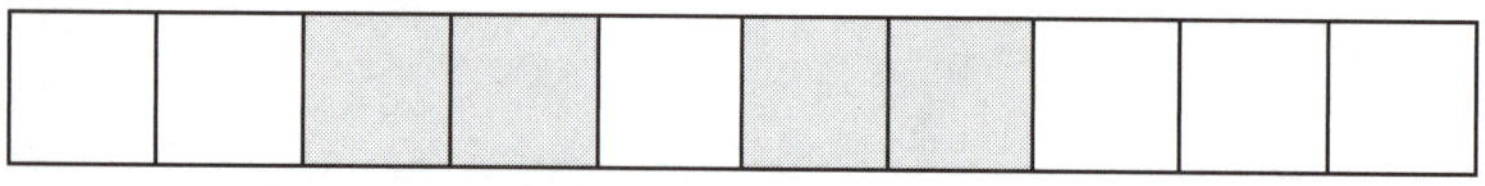

3. _____ _____

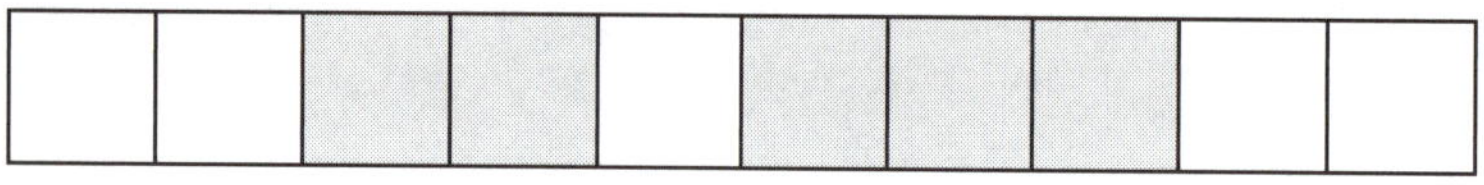

4. _____ _____

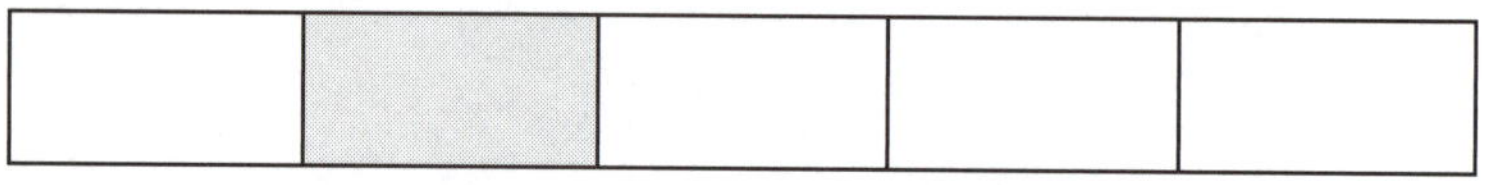

5. _____ _____

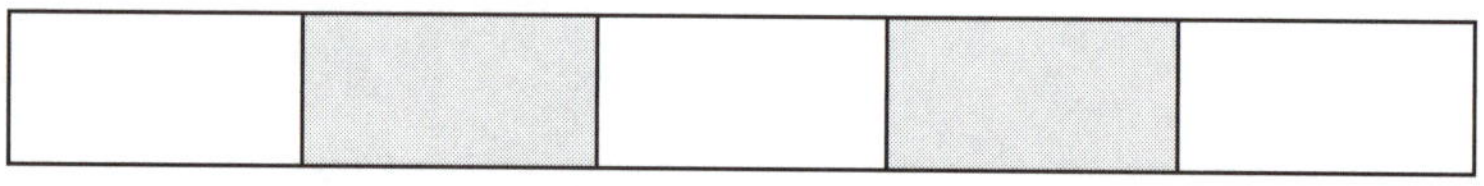

6. _____ _____

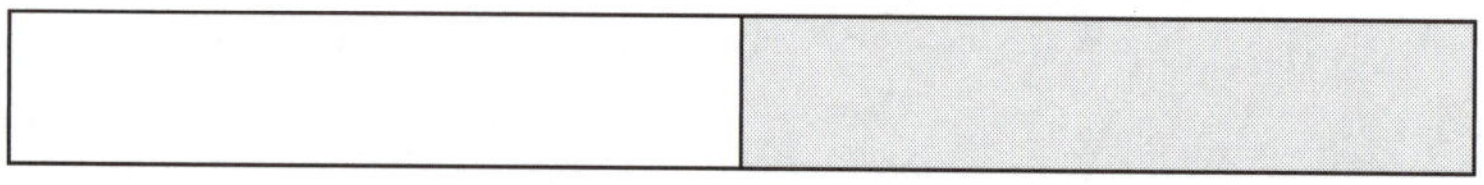

Match the numbers of the problems that have equal decimal answers.

1. _____ = _____

2. _____ = _____

3. _____ = _____

In the following problems, a flat = 1, a rod = 0.1, and a small cube = 0.01. Write the fraction and the decimal for the shaded part of each figure.

	Fraction	**Decimal**

1. 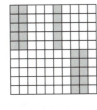   _____   _____

2. 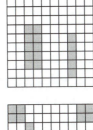   _____   _____

3. 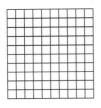   _____   _____

Shade in the 100 grid that represents a flat to show the correct fraction or decimal and fill in the missing number in each problem.

	Fraction	**Decimal**

1. 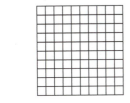 $\dfrac{63}{100}$ _____

2.   _____   0.07

3. $\dfrac{14}{100}$ _____

Activity 2: Who's First?

PURPOSE Develop an understanding of comparing and ordering decimal numbers.

MATERIALS Base Ten Blocks (page A-13 and A-15), Place-Value Operations Board (page A-29), and a colored chip

GROUPING Work individually or in pairs.

GETTING STARTED Represent the decimal point by placing the chip between the large cube (1) and the flat (0.1). Construct each of the numbers on your place-value operations board.

PLACE-VALUE OPERATIONS BOARD

Write the correct symbol (< , > , =) in the box to make each statement true.

a. 0.172 ☐ 0.34

b. 0.081 ☐ 0.28

c. 0.75 ☐ 0.57

d. 0.104 ☐ 0.2

1. In each of the following problems, place the decimals in the correct boxes to make the statement true.

   a. 0.321
      0.132
      0.44
      ☐ > ☐ > ☐

   b. 0.019
      0.91
      0.109
      ☐ < ☐ < ☐

   c. 0.230
      0.302
      0.203
      ☐ > ☐ > ☐

2. Explain the method that you used to order the decimals.

Activity 3: Race for the Flat

PURPOSE   Develop an algorithm for addition of decimals.

MATERIALS   Base Ten Blocks (page A-13) and number cubes labeled as shown

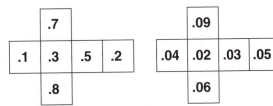

GROUPING   Work in pairs.

GETTING STARTED   In this game, the flat = 1, the rod = 0.1, and the cube = 0.01. Players alternate rolling the tenths and hundredths dice. After each roll, a player collects a number of base ten blocks equivalent to the corresponding numbers on the dice and places them on his or her flat. On each roll after the first, a player adds the new blocks to those on the flat, trading when necessary. Each addition must be recorded and validated by the blocks on the flat as shown below. The winner is the first player to cover the flat exactly. On any turn, a player may choose to roll only one die.

Example: Turn 1   Turn 2

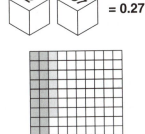

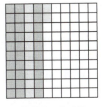

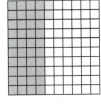

0.27   +   **0.27 + 0.15**   =   **0.42**

Play another game in which a player does not have to fill the flat exactly. Choose a goal for the game from one of the following:

a. The winner is the person whose number is closest to 1.
b. The winner is the person whose number is the greatest.

Players roll the dice and add blocks to the flat as above. For goal a, a player's strategy will determine when to stop rolling the dice. For goal b, players stop adding blocks on the roll that covers the flat, with or without some remainder.

At the conclusion of a game, players compare the decimal numbers represented by all of their blocks. The winner is the person whose number meets the goal of the game.

Activity 4: Empty the Board

PURPOSE   Develop an algorithm for subtraction of decimals using Base Ten Blocks.

MATERIALS   Base Ten Blocks (page A-13 and A-15), tenths and hundredths decimal dice (see Activity 3), a colored chip, and the Place-Value Operations Board (page A-29)

GROUPING   Work in pairs.

GETTING STARTED   In this activity, a flat = 1, a rod = 0.1, and the small cube = 0.01. Each player arranges a stack of five flats on his or her place-value operations board. Players alternate turns rolling the dice. After each roll, the player who rolled the dice removes blocks equivalent to the sum of the numbers on the dice, making trades when necessary. Then the player must record the subtraction problem that was modeled by removing the blocks. The first player to empty the entire board is the winner.

Example:

PLACE-VALUE OPERATIONS BOARD

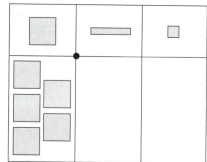

Roll 1

$$\begin{array}{r} 5.00 \\ -\ .69 \\ \hline 4.31 \end{array}$$

Roll 2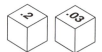

$$\begin{array}{r} 4.31 \\ -\ .23 \\ \hline 4.08 \end{array}$$

EXTENSIONS   1. Let the large cube = 1, a flat = 0.1, a rod = 0.01, and a small cube = 0.001. Also use the thousandths die.

2. How would you adapt this game to reinforce addition of decimals, and how would you adapt the game in Activity 3 for subtraction?

DECIMAL PUZZLE

In each of the given numbers, the decimal point may be placed in front of the first digit, behind the last digit, or between any two digits. Place the decimal point in each number so that the sum of the resulting numbers is 100.

$$\begin{array}{r} 6\ 8 \\ 9\ 5\ 9 \\ 4\ 8\ 1 \\ 1\ 7\ 3\ 4 \\ 2\ 2\ 7\ 9 \\ +\ 4\ 0\ 1\ 1 \\ \hline \end{array}$$

Answer is 100

Activity 5: Decimal Arrays

PURPOSE   Develop an algorithm for multiplication of decimals.

MATERIALS   Base Ten Blocks (page A-13 and A-15)

GROUPING   Work individually or in pairs.

GETTING STARTED   Recall from Activity 8 in Chapter 6 on multiplication of fractions that 0.2×0.3 means 0.2 of 0.3. In this activity, a flat = 1, a rod = 0.1, and the small cube = 0.01.

Example: To determine 0.2×0.3, place a flat on your paper and label the factors as shown.

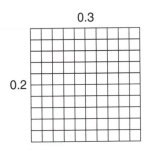

Place three rods vertically on the flat as shown to model 0.3.

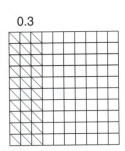

Place two rods horizontally as shown to model 0.2.

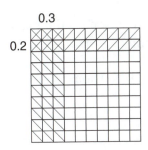

The product of 0.2×0.3 is determined by the rectangular array formed by the overlapping parts of the rods. The six squares in the array represent *6 parts of 100*.

So, 0.2 of 0.3 or $0.2 \times 0.3 = 0.06$.

Use base ten blocks to solve these multiplication problems. Record your results on the 100 grids by labeling the dimensions (factors) and shading in the correct number of rods. Find the number of squares in the overlapping area to determine each product.

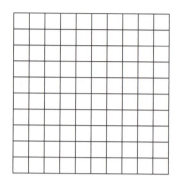 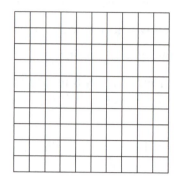 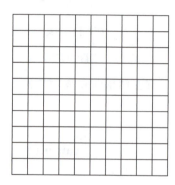

$0.4 \times 0.7 =$ _____ $0.5 \times 0.2 =$ _____ $0.8 \times 0.9 =$ _____

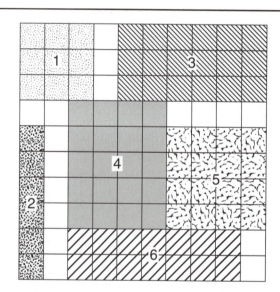

The 100 grid above represents a flat that is equal to 1. Write the dimensions for each shaded rectangle and determine its area.

Rectangle	Dimensions	Area	Rectangle	Dimensions	Area
1	____ × ____	____	2	____ × ____	____
3	____ × ____	____	4	____ × ____	____
5	____ × ____	____	6	____ × ____	____

How can you use the dimensions of the rectangle to determine its area?

Activity 6: Decimal Multiplication

PURPOSE   Use Base Ten Blocks to extend multiplication of decimals.

MATERIALS   Base Ten Blocks (page A-13 and A-15) and the Multiplication and Division Frame (page A-26)

GROUPING   Work individually.

For these problems, a flat = 1, a rod = 0.1, and a small cube = 0.01. Model the multiplication by using base ten blocks to construct a rectangle in the multiplication and division frame. Then determine the product of the two factors. Record your work as shown in the example.

Example:

$$
\begin{array}{rrr}
2 & + & .4 \\
\times \quad 1 & + & .3 \\
\hline
 & & .12 \\
 & .6 & \\
 & .4 & \\
2 & & \\
\hline
2 & + \quad 1.0 \quad + & .12
\end{array}
$$

$\rightarrow$ 12 cubes
$\rightarrow$ 6 rods
$\rightarrow$ 4 rods
$\rightarrow$ 2 flats

= **3.12**

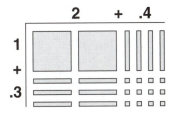

1. 3×2.3 _____

2. 0.5×3.4 _____

3. 3.4×1.2 _____

4. 0.4×5.6 _____

EXTENSIONS   Construct some rectangular arrays to represent the product of two decimals. Give the arrays to another student, identifying the correct decimal value for each Base Ten Block. Instruct the student to determine the factors of your number.

Activity 7: Target Number Revisited

PURPOSE Develop number and operation sense, estimation skills, and an increased understanding of multiplication and division of decimals.

MATERIALS A calculator for each pair of students

GROUPING Work in pairs.

GETTING STARTED Recall the Target Number game in Activity 9 in Chapter 3. All computations will be done on the calculator, but determining the guesses to enter must be done mentally.

Game 1 Player 1 chooses the target product and Player 2 chooses a factor. Then follow the steps in the flow chart.

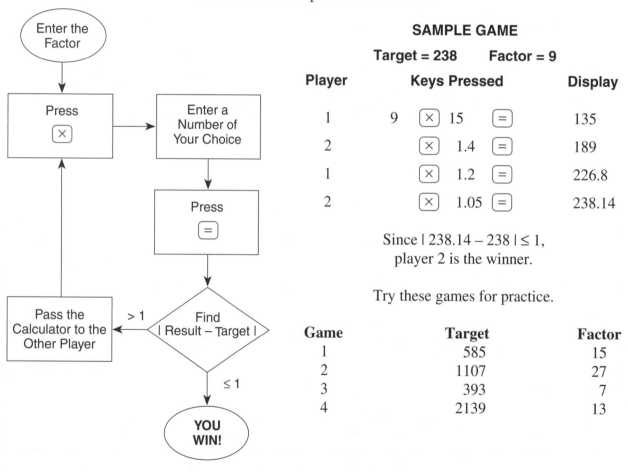

SAMPLE GAME

Target = 238   Factor = 9

Player	Keys Pressed				Display
1	9	× 15	=		135
2		× 1.4	=		189
1		× 1.2	=		226.8
2		× 1.05	=		238.14

Since | 238.14 − 238 | ≤ 1, player 2 is the winner.

Try these games for practice.

Game	Target	Factor
1	585	15
2	1107	27
3	393	7
4	2139	13

EXTENSIONS **Game 2** Player 1 chooses a target quotient, and Player 2 chooses a product. Follow the steps in the flow chart, but enter the product and use the ÷ key instead of the × key. *The winner is the player who gets within* 0.5 *of the target quotient.*

Activity 8: What Is Percent?

PURPOSE   Develop the concept of percent as a part of 100 and the relationship among fractions, decimals, and percents.

MATERIALS   A transparent copy of a 100 Grid (page A-37)

GROUPING   Work individually.

GETTING STARTED   Note that the % symbol is a special arrangement of a 1 and two 0s.

Percent means *part of* 100.

Example: 40% means 40 equal parts out of 100.

80 correct out of 100 problems $= \dfrac{80}{100} = 80\%$.

Place a copy of the 100 grid over each square. Count the shaded squares in the grid to determine what part of 100 and the percent of each figure that is shaded.

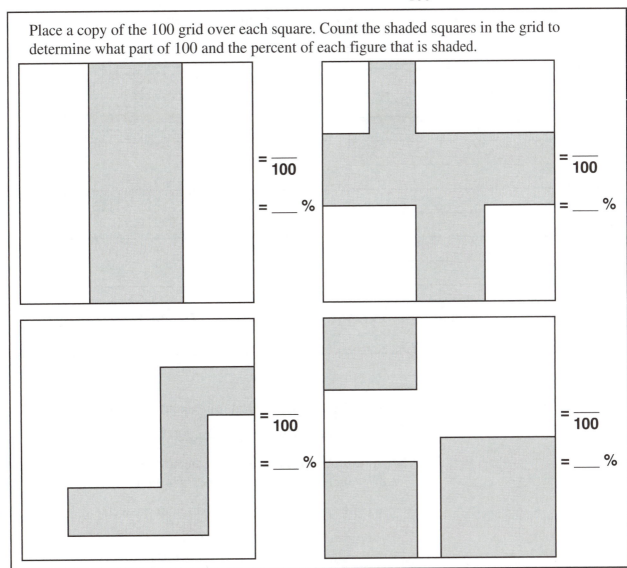

Name the fractional part of each figure that is shaded. Use your 100 grid to determine the equivalent decimal and percent.

1.

_____ = _____ = _____
Fraction   Decimal   Percent

2.

_____ = _____ = _____
Fraction   Decimal   Percent

3.

_____ = _____ = _____
Fraction   Decimal   Percent

4.

_____ = _____ = _____
Fraction   Decimal   Percent

EXTENSIONS

1. List five real-world applications of percent between 28% and 57%.

2. List three real-world applications of 200%.

3. A headline on the business page of the newspaper reads, *ABC stock price increases 200% in the past six months from 7 1/2 to 15.*

   Do you believe the headline? Explain your answer.

Activity 9: What Does It Mean?

PURPOSE   Reinforce the relationships among fractions, ratios, and percents.

MATERIALS   A transparent copy of a 100 Grid (page A-37)

GROUPING   Work individually.

Complete the following.

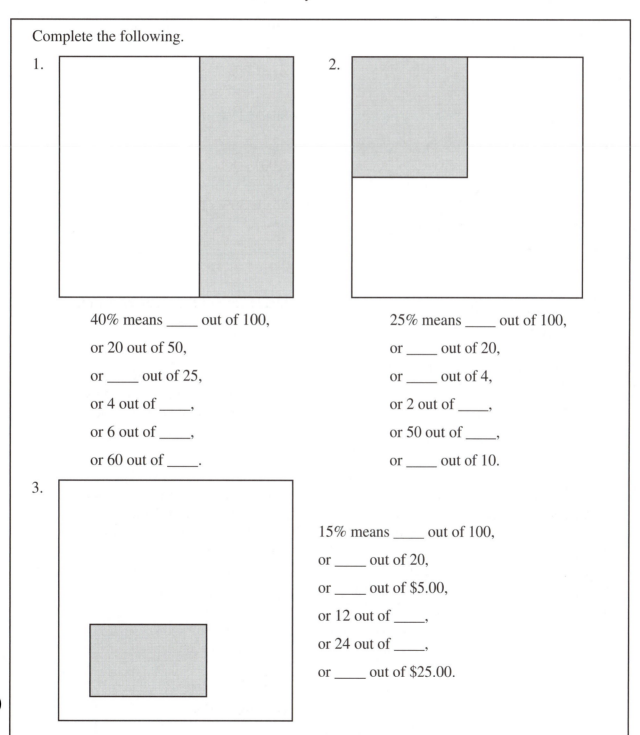

1.

40% means _____ out of 100,

or 20 out of 50,

or _____ out of 25,

or 4 out of _____,

or 6 out of _____,

or 60 out of _____.

2.

25% means _____ out of 100,

or _____ out of 20,

or _____ out of 4,

or 2 out of _____,

or 50 out of _____,

or _____ out of 10.

3.

15% means _____ out of 100,

or _____ out of 20,

or _____ out of $5.00,

or 12 out of _____,

or 24 out of _____,

or _____ out of $25.00.

Determine the number of figures to be shaded in each problem.

1. Shade 20% of the squares.

$$\frac{20}{100} = \frac{}{15}$$

2. Shade 75% of the triangles.

$$\frac{75}{100} = \frac{}{8}$$

3. Shade 25% of the rectangles.

$$\frac{25}{100} = \frac{}{8}$$

4. Shade 30% of the circles.

5. Shade 25% of the triangles. Circle 50% of the triangles.

6. Put an X in 35% of the rectangles.
   Put a Y in 25% of the rectangles.
   Put a Z in 15% of the rectangles.

7. Put an A in 10% of the squares.
   Put a B in 27% of the squares.
   Put a C in 18% of the squares.
   Put a D in 35% of the squares.
   Put an E in 3% of the squares.

   How many A's? _____

   How many B's? _____

   How many C's? _____

   How many D's? _____

   How many E's? _____

Activity 10: Flex-It

PURPOSE   Reinforce the concept of percent and apply the concept of percent increase to measurements of the human body.

MATERIALS   A centimeter tape measure and a calculator

GROUPING   Work in groups of 4 or 5.

GETTING STARTED   Enter the names of the people in your group in the table below. Predict the order (largest to smallest) for the percent increase for each muscle for each member of the group and enter your guesses in the table. The highest rating is 1.

Make the following measurements in centimeters and round to the nearest 0.5 cm.

A. Extend an arm. Measure the largest circumference of the biceps while it is relaxed.

B. Extend a leg. Measure the largest circumference of the calf while it is relaxed.

C. Flex the arm and leg as much as possible. Measure the largest circumference of the biceps and the calf muscles.

Determine the percent increase for each muscle and record it in the table.

Name	Arm Extended	Arm Flexed	Percent Increase	Rank Order of Increase Est.	Act.	Leg Extended	Leg Flexed	Percent Increase	Rank Order of Increase Est.	Act.

1. How does the final order of the percent increases compare with your predictions?

2. Did the most muscular person in your group have the greatest percent increase? _____ Explain.

3. Explain how a person's size and body structure affect the percent increase in the size of the muscles measured in this activity.

Chapter 6 Summary

The activities in this chapter stressed the conceptual development of decimals. The activities continued the constructivist design of Chapter 6. Models were used to demonstrate the relationship between fractions and decimals and the activities progressed from the concrete to the representational to the abstract or symbolic level.

The game format used in Activities 3 and 4 provided the opportunity to physically place rods and cubes on a flat during the addition and subtraction process and then to make the necessary trades in order to represent the answer with the least number of blocks. This physical modeling of the operations and the trading and regrouping process are all essential for understanding the computational algorithms.

Activities 5 and 6 used the rectangular array model for multiplication to extend the processes developed with whole numbers and fractions to decimals. The area model will be revisited in the geometry section. Using this model in a variety of settings provides a firm foundation for understanding the concept of area.

Activity 7 revisited the Target Number game from Chapter 3. The guess and check strategy was used with the calculator to reinforce number and operation sense, and estimation skills. This activity also developed an increased understanding of multiplication and division of decimal numbers.

In Activity 8, the relation between the percent symbol, %, and *part of 100* was explained. The use of the 100 Grid in Activities 8 and 9 established the relationship between percent and 100 and also helped to develop the connections among fractions, decimals, ratios, and percents. These connections were applied in determining the percent of a number by finding a fractional part or multiplying by a decimal.

Chapter 7
Probability

"A subject in its own right, probability is connected to other areas of mathematics, especially number and geometry. Ideas from probability serve as a foundation to the collection, description, and interpretation of data.

In prekindergarten through grade 2, the treatment of probability ideas should be informal. Teachers should build on children's developing vocabulary to introduce and highlight probability notions, for example, "We'll *probably* have recess this afternoon," or "It's *unlikely* to rain today." Young children can begin building an understanding of chance and randomness by doing experiments with concrete objects, such as choosing colored chips from a bag. In grades 3–5 students can consider ideas of chance through experiments—using coins, dice, or spinners—with known theoretical outcomes or through designating familiar events as impossible, unlikely, likely, or certain. Middle-grades students should learn and use appropriate terminology and should be able to compute probabilities for simple compound events, such as the number of expected occurrences of two heads when two coins are tossed 100 times. In high school, students should compute probabilities of compound events and understand conditional and independent events. Through the grades, students should be able to move from situations for which the probability of an event can readily be determined to situations in which sampling and simulations help them quantify the likelihood of an uncertain outcome."
                    —*Principles and Standards for School Mathematics*

The activities in this chapter develop the basic ideas and vocabulary of probability and introduce several different models used to determine probabilities. You will learn to use ratios to assign a probability to an event and to use and interpret both experimental and theoretical probabilities. In the process, you will develop an understanding of fair and unfair games and random events.

Most people have an intuitive understanding of probability even though they have had limited formal educational experiences with it. The goal of this chapter is to extend your intuitive ideas to a sound mathematical understanding of probability.

Activity 1: What Are the Chances?

PURPOSE   Introduce the concept of probability using intuitive ideas about chance.

GROUPING   Work individually.

GETTING STARTED   We frequently encounter situations in which we cannot predict the outcome in advance. In such situations, we often talk about the chances of an outcome occurring. If we think the chances that something will happen are good, we might say it is *likely* or it is probable. On the other hand, when we think the chances are poor, we often say the outcome is *unlikely* or not very probable.

A paper bag contains eight marbles: five green, two blue, and one yellow. Suppose one marble is drawn from the bag. Use the scale below to describe the chances of each of the following events occurring. Explain your decision for each event.

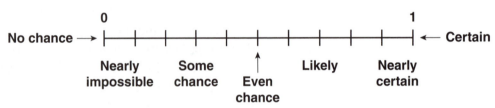

1. A green marble is drawn.
2. A blue marble is drawn.
3. A yellow marble is drawn.
4. A red marble is drawn.
5. A blue or yellow marble is drawn.
6. The marble drawn is not yellow.

Make a spinner face by dividing each circle into three sections and labeling each section with a color (RED, WHITE, or BLUE) so that the given condition is true.

1. The spinner is certain to stop on RED.

2. The spinner is likely to stop on WHITE.

3. The spinner can't stop on BLUE.

4. There is little chance the spinner will stop on RED.

5. The spinner will probably stop on RED or BLUE.

6. The spinner has the same chance of stopping on RED, WHITE, or BLUE.

Activity 2: The Spinner Game

PURPOSE   Develop and extend the concept of probability and introduce the idea of a fair game.

MATERIALS   A #1 paper clip (approximately $1\frac{5}{16}$ in. long) for each student

GROUPING   Work in pairs.

GETTING STARTED   Make a spinner by bending a paper clip into the shape shown below. The long, straight part will be the pointer. Place your pencil through the loop of the paper clip and put the point of the pencil on the center of the spinner. Spin the spinner by flicking the paper clip with your finger.

Rules for the Spinner Game:

• This is a game for two players. One player spins Spinner A and the other spins Spinner B.
• Both players spin their spinner at the same time. The player who spins the greatest number is the winner.

Spinner A

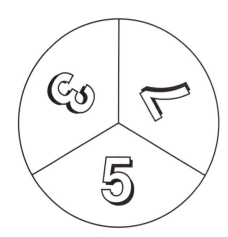

Spinner B

Play the game 20 times. Record the winner of each game in the table. You may not change spinners once the game has begun.

Do you think this is a fair game? Explain.

Winning Spinner		
Spinner	**Tally**	**Frequency**
A		
B		
Total		

Combine your data for the Spinner Game with the data from the other teams in your class to find a class total for the number of wins for Spinner A and Spinner B. Record the data in the table.

Does this data change your opinion about whether the game is fair? Explain.

Winning Spinner	
Spinner	Frequency
A	
B	
Total	

EXTENSIONS   Play the following spinner game.

Rules:
- This is a game for two players.
- The first player chooses a spinner from Spinners A, B (preceding page), or C (below).
- The second player chooses a spinner from the two remaining spinners.
- Both players spin at the same time. The player who spins the greatest number is the winner.

Spinner C

Play the game using different combinations of spinners. Then answer the following questions.

1. How does allowing the first player to choose a spinner affect the fairness of this game?

2. Is there a strategy for choosing the spinners that would give one player an advantage over the other? Explain.

Activity 3: Numbers That Predict

PURPOSE   Introduce experimental and theoretical probability and mutually exclusive, complementary, equally likely, certain, and impossible events.

GROUPING   Work individually or in pairs.

GETTING STARTED   A meteorologist says the chance of snow today is 20%. The fine print in the sweepstakes announcement indicates that your odds of winning are one in 2.8 million. In a group of 25 nineteen-year-old women, about 22 will never have been married. These are just three examples of situations in which numbers are used to predict.

EXPERIMENTAL PROBABILITY

1. Spin Spinner A from Activity 2 thirty times and record the results in the table.

A *probability* is a ratio that predicts the chance or likelihood of something happening.

Outcome	Tally	Frequency
2		
4		
9		
Total		

$$\text{Experimental probability} = \frac{\text{number of times the event occurs}}{\text{total number of trials}}$$

2. Use the data in the table to calculate the following experimental probabilities. [$P(2)$ = the probability the spinner stops in the region labeled 2.]

$P(2) =$ _____ $P(4) =$ _____ $P(9) =$ _____

$P(\text{square number}) =$ _____ $P(\text{odd number}) =$ _____ $P(\text{even number}) =$ _____

3. What is $P(1)$? Why?

4. What is $P(\text{a number less than 10})$? Why?

5. a.   What does it mean for an event to have a probability of zero?

    b.   a probability of one?

COMPLEMENTARY EVENTS

1. What do you observe about P(even number) + P(odd number)?

2. If the spinner does not stop on an even number, then it must stop on an odd number. For this reason, the events "the spinner stops on an even number" and "the spinner stops on an odd number" are called *complementary events*. What can you conclude about the probabilities of complementary events?

THEORETICAL PROBABILITY

When you spin Spinner A, there are three possible outcomes. Since each of the central angles on the spinner has the same measure, over many trials, each of the outcomes should occur about the same number of times. That is, the outcomes are *equally likely*.

If the outcomes of an experiment are equally likely, the theoretical probability of an event may be calculated without conducting an experiment.

$$\text{Theoretical probability} = \frac{\text{number of outcomes making up the event}}{\text{total number of possible outcomes}}$$

1. Calculate the following theoretical probabilities for spinning Spinner A.

   $P(2) =$ _____ $P(4) =$ _____ $P(9) =$ _____

   P(square number) = _____ P(odd number) = _____ P(even number) = _____

2. Compare the probabilities in Exercise 1 above to the experimental probabilities you calculated in Exercise 2 on the preceding page. Explain any similarities or differences.

MUTUALLY EXCLUSIVE EVENTS

1. a. What is P(2 or 4)?

    b. The events "the spinner stops in the region labeled 2" and "the spinner stops in the region labeled 4" cannot happen at the same time. Events that cannot occur simultaneously are said to be *mutually exclusive*. What is P(2) + P(4)?

    c. What is P(odd number or square number)?

    d. What is P(odd number) + P(square number)?

    e. Are the events "the spinner stops on an odd number" and "the spinner stops on a square number" mutually exclusive? Explain.

    f. Based on your answers to Parts a–e, if A and B are mutually exclusive events, what is P(A or B)?

2. What is the difference between mutually exclusive events and complementary events?

EXTENSIONS

1. Design a spinner that has two regions, one labeled **red** and the other **blue**, and such that $P(\text{red}) = \frac{1}{4}$.

2. Spin the spinner 20 times and record the number of times it stops on red. Use the results to calculate the following experimental probabilities.

    $P(\text{red}) = $ _____ $P(\text{blue}) = $ _____

3. Did the probabilities come out exactly as you expected? Explain.

4. Combine the results of your 20 spins with those of four classmates. Calculate the experimental probabilities for the 100 spins. What do you observe?

Activity 4: It's in the Bag

PURPOSE	Determine theoretical and experimental probabilities and use random sampling and probabilities to make predictions.
MATERIALS	A paper bag and eight green, eight yellow, and eight blues quares
GROUPING	Work in pairs.

1. A paper bag contains five green squares, two blue squares, and one yellow square. If one square is drawn from the bag, what is the theoretical probability of each event below?

   a. A green square is drawn.   b. A blue square is drawn.

   c. A yellow square is drawn.   d. A red square is drawn.

   e. A blue or yellow square is drawn.   f. The square drawn is not yellow.

2. a. Plot the probabilities from Exercise 1 on the following number line.

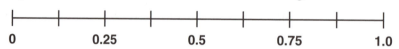

   b. For each event, select the phrase that most accurately describes the chance of it occurring: *no chance, nearly impossible, some chance, an even chance, likely, nearly certain,* or *certain.*

3. How many blue squares must be added to the bag in Exercise 1 so that $P(\text{blue}) = 0.5$? Explain how you determined your answer.

4. Simulate the experiment in Exercise 1. Put five green squares, two blue squares, and one yellow square in a paper bag. Shake the bag and draw one square. Record the color on a tally sheet and put the square back in the bag. Repeat the experiment until you have completed 40 trials.

5. Use the results from the simulation in Exercise 4 to find the following experimental probabilities.

   a. $P(\text{green})$ b. $P(\text{blue})$ c. $P(\text{yellow})$

   d. $P(\text{red})$ e. $P(\text{blue or yellow})$ f. $P(\text{not yellow})$

6. Compare the experimental probabilities to the theoretical probabilities in Exercise 1.

7. a. Make up a bag containing nine squares of three different colors. Exchange bags with your partner.

   b. Repeat the sampling procedure in Exercise 4 until you have completed 45 trials.

   c. Predict how many squares of each color are in the bag. Then check your prediction by looking in the bag.

   d. How could you use experimental probabilities to help make your predictions?

Activity 5: Paper-Scissors-Rock

PURPOSE	Introduce the use of a matrix and a tree diagram to calculate theoretical probabilities, and reinforce fair games.
GROUPING	Work in pairs.
GETTING STARTED	The Paper-Scissors-Rock game has been popular for many years. The two-player game is played as follows:

- Each player makes a fist.
- On the count of three, each player shows either *scissors* by showing two fingers, *paper* by showing four fingers, or *rock* by showing a fist.
- If scissors and paper are shown, the player showing scissors wins, since the scissors cut paper.
- If scissors and rock are shown, the player showing rock wins, since a rock breaks the scissors.
- If paper and rock are shown, the player showing paper wins, since paper wraps a rock.

---

1. Do you think Paper-Scissors-Rock is a fair game? Explain.

2. Play the game 45 times. Each player should tally the outcomes in a table like the one below.

		Your Partner		
		Paper	**Scissors**	**Rock**
You	**Paper**			
	Scissors			
	Rock			

3. Use the data in your table to calculate the following experimental probabilities.

   P(you win) = _____ P(your partner wins) = _____ P(tie) = _____

   P(you show rock) = _____ P(you show paper and your partner shows rock) = _____

4. Use the probabilities in Exercise 3 to decide whether Paper-Scissors-Rock is a fair game. Explain your decision.

PROBABILITIES USING A MATRIX

You can determine if the game is fair without conducting an experiment.

1. Complete the *matrix* at the right.

2. If the players choose the sign they show randomly, each of the nine outcomes in the matrix are *equally likely*. Find the following probabilities in this case.

   A = A wins
   B = B wins
   T = Tie

   P(A wins) = _____

   P(B wins) = _____

   P(Tie) = _____

3. Based on the probabilities in Exercise 2, is this a fair game? Explain.

		Paper	Scissors	Rock
Player A	Paper			A
	Scissors		T	
	Rock			

(Player B)

PROBABILITIES USING A TREE DIAGRAM

Since Paper-Scissors-Rock can be thought of as a multi-stage experiment, it can be analyzed using a *tree diagram*.

1. Complete the tree diagram at the right.

2. What does the outcome PP in the tree diagram mean?

Consider the path leading to the outcome **PP**. Since the choices made by Player A and Player B are independent of one another, based on the probabilities along the path, we would expect the following:

- In $\frac{1}{3}$ of the games played, Player A will show paper.

- Player B will show paper in $\frac{1}{3}$ of the games in which Player A shows paper $\left(\frac{1}{3} \text{ of } \frac{1}{3}\right)$.

This shows that the probability of each outcome is the product of the probabilities along the path leading to the outcome.

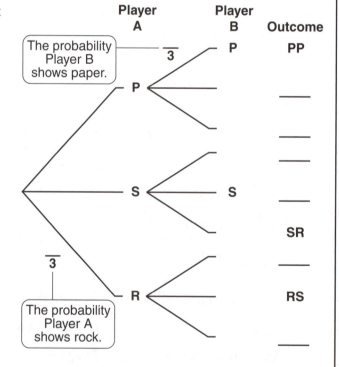

Use the probabilities in the tree diagram to find the following theoretical probabilities.

1. $P(PS) = $ _____

2. $P(SP \text{ or } SR) = $ _____

3. $P(A \text{ wins}) = $ _____

4. $P(B \text{ wins}) = $ _____

5. $P(\text{at least one player shows scissors}) = $ _____

THE SPINNER GAME REVISITED

1. Use your individual data from Activity 2 to calculate the following experimental probabilities:

   $P(\text{spinner A wins}) = $ _____ $P(\text{spinner B wins}) = $ _____

2. Use your class totals to calculate the following experimental probabilities:

   $P(\text{spinner A wins}) = $ _____ $P(\text{spinner B wins}) = $ _____

3. a. Construct a matrix for the Spinner Game in Activity 2.

   b. Are the outcomes in the matrix equally likely? Explain.

4. Use the data in the matrix to calculate the following theoretical probabilities:

   $P(\text{spinner A wins}) = $ _____ $P(\text{spinner B wins}) = $ _____

5. a. Compare the probabilities in Exercise 4 to those in Exercise 1.

   b. Compare the probabilities in Exercise 4 to those in Exercise 2.

6. Is the spinner game a fair game? Explain.

EXTENSIONS   Do the extensions in Activity 2, but instead of playing the game, analyze it using matrices.

Activity 6: What's the Distribution?

PURPOSE	Determine experimental probabilities and introduce probability distributions.
MATERIALS	Six pennies, five dice, and one paper cup per student; colored pencils
GROUPING	Work in pairs.

1. If you place all six coins in the paper cup, shake it thoroughly, and spill the coins on a table, do you think you will get one combination of heads and tails more often than the others? If so, what combination will it be and why?

2. Do you think you will get some combinations of heads and tails less often than others? Explain.

3. Perform the experiment of spilling the coins 35 times. After each spill, record the number of heads by coloring in a square in the row corresponding to the number of heads in the grid below.

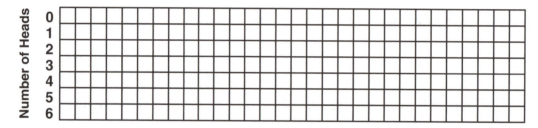

4. Compare the results with your predictions in Exercises 1 and 2.

5. Use a different colored pencil to add the results of your partner's 35 trials to the grid. Use the data for the 70 trials to calculate the following experimental probabilities.

    $P(0 \text{ heads}) = $ _____ $P(1 \text{ head}) = $ _____ $P(2 \text{ heads}) = $ _____

    $P(3 \text{ heads}) = $ _____ $P(4 \text{ heads}) = $ _____ $P(5 \text{ heads}) = $ _____

    $P(6 \text{ heads}) = $ _____

6. Compare your combined results with that of another team. What similarities and differences do you notice in

    a. the distribution of the outcomes in Exercise 3?

    b. the probabilities in Exercise 5?

1. If you do the experiment again using five coins instead of six, how will the distribution of heads change? Why?

2. Use five coins. Perform the experiment 35 times. Record the results in the grid below.

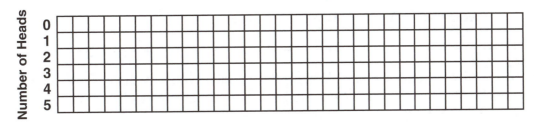

3. Were your predictions in Exercise 1 correct? Explain.

4. Use the data for the 35 trials to calculate the following experimental probabilities.

    P(0 heads) = _____ P(1 head) = _____ P(2 heads) = _____

    P(3 heads) = _____ P(4 heads) = _____ P(5 heads) = _____

BIASED COINS

How would the distribution of heads in the experiment of spilling five coins from a cup change if the probability of a coin showing heads was $\frac{1}{3}$ instead of $\frac{1}{2}$? To find out, try the following experiment.

1. Use five dice in place of the five coins. Spill the dice from the cup. A die that lands with a 1 or 2 facing up is counted as a head. One that lands with a 3, 4, 5, or 6 facing up is counted as a tail. Repeat the experiment 35 times and record the number of heads in a grid as above.

2. Describe the similarities and differences between the two distributions and explain the reasons for them.

EXTENSIONS   Repeat the biased coin experiment for coins that have a probability of $\frac{5}{6}$ of showing heads.

Activity 7: Pascal's Probabilities

PURPOSE   Find theoretical probabilities in situations involving independent events using tree diagrams and Pascal's Triangle.

GROUPING   Work individually or in pairs.

Suppose a family has three children. What is the probability that all three children are girls? Two are girls? Only one is a girl? None are girls? A tree diagram can be used to determine the theoretical probabilities of each number of girls.

1st Child	2nd Child	3rd Child	Outcome	Number of Girls
		G	GGG	3
	G	B	GGB	2
G		G	GBG	2
	B	B	GBB	1
		G	BGG	2
	G	B	BGB	1
B		G	BBG	1
	B	B	BBB	0

Key: G = girl
      B = boy

1. Assuming that when a child is born the probability that it is a girl is 0.5, are the outcomes shown above equally likely? Explain.

2. Complete the following to summarize the results from the tree diagram.

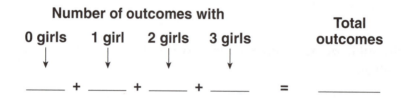

Number of outcomes with

0 girls   1 girl   2 girls   3 girls   **Total outcomes**

↓   ↓   ↓   ↓

_____ + _____ + _____ + _____   =   _____

3. Use the results above to determine the following theoretical probabilities for the 3-child family.

$P(0 \text{ girls}) =$ _____ $P(1 \text{ girl}) =$ _____

$P(2 \text{ girls}) =$ _____ $P(3 \text{ girls}) =$ _____

Tree diagrams can also be used to find the outcomes for families with other numbers of children, but such diagrams quickly become very large and complex. The results for families with 1, 2, 3, 4, and 5 children are shown in the diagram below. The results are listed in order from the number of outcomes with 0 girls to the number with all girls.

					□							
1 Child →					1		1					
2 Children →				1		2		1				
3 Children →			1		3		3		1			
4 Children →		1		4		6		4		1		
5 Children →	1		5		10		10		5		1	
6 Children →	___	___	___	___	___	___	___					

1. Make a tree diagram to check the results for a family with four children.

2. The array of numbers above is part of Pascal's Triangle, named in honor of Blaise Pascal, a seventeenth-century French mathematician and philosopher who was one of the founders of probability theory. Look for patterns in the array to help complete the following.

    a. What number should be placed in the box in the first row of the triangle?

    b. How can the numbers in each row of Pascal's Triangle be determined from the numbers in the preceding row?

    c. Complete the seventh row of Pascal's Triangle.

3. a. Use the numbers in the sixth row of Pascal's Triangle to calculate the theoretical probability of each number of girls in a family with five children.

    b. How do the theoretical probabilities compare to the experimental probabilities for the numbers of heads when five coins are spilled from a cup in Activity 6? Explain any similarities or differences.

4. Use the numbers in the seventh row to help answer the following questions. What is the probability that in a family with six children

    a. there are exactly three girls?   b. there are at least three girls?

    c. there is exactly one boy?   d. there are more than two boys?

5. Use Pascal's Triangle to calculate the theoretical probability for each number of girls in a family with nine children. Explain your procedure.

Activity 8: Simulate It

PURPOSE   Analyze probability situations using simulations.

MATERIALS   One die

GROUPING   Work individually or in pairs.

DESIGNING A SIMULATION

Robin Hood and Maid Marian are having an archery contest. They alternate turns shooting at a target. The first person to hit it wins. The probability that Robin hits the target is $\frac{1}{3}$ and the probability that Marian hits it is $\frac{2}{5}$. Since Marian is the better archer, Robin shoots first. What is the probability that Marian will win the contest? On average, how many shots will be necessary to determine a winner? To find the answers, you can simulate the problem.

Step 1: Select a model.

- To simulate Robin's shot, roll a die. If the result is a 1 or 2, he hits the target. Otherwise he misses.
- For Marian's shot, roll a die. If the outcome is a 1 or 2, she hits the target. If it is a 3, 4, or 5, she misses. If it is a 6, roll the die again.

Step 2: Conduct a trial and record the result.

A trial consists of rolling a die to alternately simulate a shot for Robin and then one for Marian until someone hits the target.

Example: Shot	Archer	Die Roll	Result
1	Robin	4	Miss
2	Marian	6	Roll Again
		5	Miss
3	Robin	1	Hit—Robin Wins!

Step 3: Repeat Step 2 until the desired number of trials is completed.

Complete 30 trials and record the results in a table like the one below.

Trial	Winner	Number of Shots	Trial	Winner	Number of Shots
Example	Robin	3			

Step 4: Interpret the results.

Based on the results of the 30 trials, what is the probability that Robin wins the contest?

What is the average number of shots for the 30 trials?

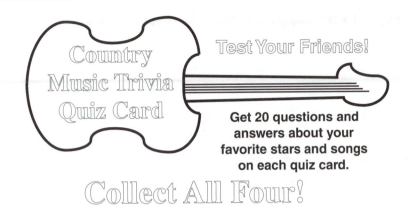

Get 20 questions and answers about your favorite stars and songs on each quiz card.

As part of a promotional campaign, a cereal manufacturer packages one Country Music Trivia Quiz Card inside each specially labeled box of cereal.

A country music aficionado wants to collect a complete set of the Country Music Trivia Quiz Cards. He wants to know how many boxes of cereal he must purchase in order to collect all four Quiz Cards.

1. Describe how the card obtained by purchasing one box of cereal can be modeled by rolling a die.

2. What would make up a trial?

3. Complete 20 trials and record the results in a table like the one below.

Trial	Number of Boxes	Trial	Number of Boxes

4. On average, how many boxes of cereal will the aficionado have to purchase in order to collect a complete set of four Country Music Trivia Quiz Cards?

5. Describe a model that could be used if one card was twice as likely to be in a box of cereal as the others.

6. In this case, what would make up a trial?

EXTENSIONS Suppose Robin and Marian change the rules of their tournament so that the winner is the person who wins two out of three contests. Draw a tree diagram to show the possible outcomes of the tournament. Use the results from the archery simulation to assign probabilities to the branches and determine the probability that Marian wins the tournament.

Activity 9: The Mystery Cube

PURPOSE	Reinforce the concept of expected value and the process of making predictions based on a sample.
MATERIALS	Three blank cubes per student (paper squares and a bag may be used in place of the cubes) and an inflatable model of the earth (a globe)
GROUPING	Work in pairs.

Write one of the digits 1, 2, 3, 4, 5, or 6 on each face of a cube. It is not necessary to use all six digits; you may write the same digit on different faces. **Do not show the cube to your partner.** (If paper squares and a bag are used, write one digit on each of six paper squares and place them in the bag.)

Roll the cube **without letting your partner see it** and tell your partner the digit on the top face. Continue rolling the cube and announcing the outcome until your partner decides to guess how many faces of the cube are labeled with each digit. Be sure to keep track of the number of times you roll the cube. (If paper squares are used, draw a square from the bag and read the number written on it. Put the square back in the bag and shake the bag thoroughly before making the next draw.)

If your partner correctly identifies how many faces are labeled with each digit, record the number of rolls in the table below. If the labels are not guessed correctly, tell your partner the number of digits that are correct and then continue rolling the cube until your partner can correctly identify how many faces are labeled with each digit.

Trial	1	2	3	4	5	6
Number of Rolls						

When your partner has identified the labels correctly, switch roles. Your partner will write digits on the faces of a cube and roll it while you try to guess how the faces are labeled. Repeat this procedure until all 6 cubes have been used.

Example: Labels on the Cube

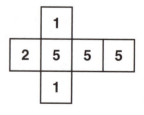

Roll: 1 2 3 4 5 6
Outcome: 5 1 5 5 1 5

Guess: There is a 1 on two faces and a 5 on four faces.
Response: You have one of the digits correct.
            (The number of faces labeled with a 1 is correct.)

Roll: 7 8 9 10
Outcome: 2 5 1 5

Guess: There is a 1 on two faces, a 2 on one face, and a 5 on three faces.
Response: That is correct.

Enter 10 rolls in the table and switch roles.

1. What was the average number of rolls needed to determine how many faces of the cube were labeled with each digit?

2. Which labeling was the most difficult to guess? Why?

3. Explain how you used (or could have used) probabilities to make your predictions.

The eight sides of an octahedral die are congruent equilateral triangles. Each side is painted either yellow, green, or blue. The results of rolling the die 240 times are summarized in the table below. Predict the number of yellow sides, the number of green sides, and the number of blue sides on the die. Explain how you arrived at your predictions.

COLOR	FREQUENCY
Green	58
Blue	27
Yellow	155
Total	**240**

EXTENSIONS   Appoint one person in the class to be the recorder. Toss an inflated globe from person to person in the classroom. When a person catches the globe, she or he should note whether her or his right index finger is touching land or water, tell the recorder, and then toss the globe to another person. The recorder tallies how many times the globe was caught with the right index finger on land and on water after 10 catches, 25 catches, and 50 catches.

1. Estimate the percent of the earth's surface that is covered by water based on

   a. 10 catches.

   b. 25 catches.

   c. 50 catches.

2. Which estimate in Exercise 1 do you think is the most accurate? Why?

3. Look up in an almanac the percent of the earth's surface that is covered by water. Compare this figure to your estimates and explain the reasons for any similarities or differences.

Activity 10: How Many Arrangements?

PURPOSE	Introduce permutations and combinations.
MATERIALS	Colored squares (page A-7) or chips
GROUPING	Work individually or in pairs.
GETTING STARTED	One square can be arranged in a row in one way.

Two different color squares can be arranged in a row in two distinct ways.

and

1. a. Use three different color squares. How many distinct ways can you arrange these three squares in a row? Record each arrangement as you make it. Enter the total number in the table.

Number of Squares	1	2	3	4
Number of Arrangements	1	2		

   b. How do you know you have found all the possible ways to arrange the squares?

   c. How is the number of arrangements of three squares related to the number of arrangements of two squares?

2. a. How many distinct ways do you think four different color squares can be arranged in a row? Why?

   b. Use four different color squares. Make as many distinct arrangements as you can with the squares. Record each arrangement as you make it.

   c. Enter the total number of arrangements in the table. How does this compare with your prediction in Part a? How is it related to the number of arrangements of three squares?

3. If you had n different color squares, how many distinct arrangements could you make? Justify your answer.

ARRANGEMENTS WITH LIKE SQUARES

1. a. Use two red squares and one green square. How many distinct ways can you arrange these three squares in a row? Record each arrangement as you make it.

    b. If all the squares in Part a had been different colors, how many arrangements would have been possible?

    c. The number of arrangements in Part a is what fraction of the number in Part b? How is this fraction related to the number of red squares?

    d. Explain why the number of arrangements of two red squares and one green square should be this fraction of the number of arrangements of three squares.

2. a. Repeat Exercise 1 Parts a–c using three red squares and one green square.

    b. If you started with 4 red squares and 1 green square, how many distinct ways do you think the five squares could be arranged in a row? Show how you made your prediction.

    c. Use four red squares and one green square. Make as many distinct arrangements as you can with the squares. How many arrangements did you find? How does this compare with the number you predicted in Part b?

3. How many distinct ways can you arrange each of the following sets of squares in a row?

    a. 2 red, 1 green, 1 blue   b. 2 red, 2 green   c. 3 red, 2 green

4. a. The number of arrangements of each set of squares in Exercise 3 is what fraction of the number of arrangements that would be possible if all of the squares had been different colors?

    b. How is each fraction in Part a related to the numbers of each color square in the set?

5. Suppose you have a set of squares some of which have the same colors. Explain how to find the number of distinct ways the squares can be arranged in a row.

In the preceding exercises, you have been working with permutations. A *permutation* is an arrangement of a group of items in which the order is important.

1. a. Suppose you have one red, one blue, one green, and one yellow square. Complete the tree diagram below to find the number of permutations that can be formed by choosing two of the squares and arranging them in a row.

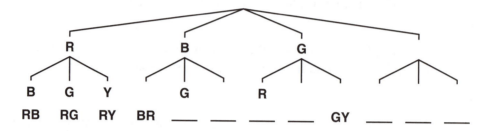

    b. How could you predict the number of permutations in Part a without using a tree diagram or listing them?

    c. How many permutations can be formed by choosing three of the squares? Explain how to find the number of permutations without listing the arrangements.

    d. Make a tree diagram to check your answer in Part c.

2. Now, suppose you want to find how many different pairs of squares can be chosen from a set containing one red, one blue, one green, and one yellow square.

    a. Is the pair consisting of the red square and the blue square (the pair RB) different from the pair BR? Explain.

    b. Does changing the order of the squares in a pair change the pair itself?

A selection of items in which order is not important is a *combination*.

3. a. For each arrangement in the tree diagram in Exercise 1 on p. 156 Part a, cross out all the other arrangements that contain the same pair of squares. How many different combinations remain?

    b. The number of combinations is what fraction of the number of permutations?

    c. How is the fraction related to the number of squares in a pair?

4. Suppose you want to choose a group of three squares from a set containing one red, one blue, one green, and one yellow square.

    a. Use your tree diagram from Exercise 1 Part d. Cross out the duplicate arrangements.

    b. How many different combinations are possible in this situation?

    c. What fraction of the number of permutations is this?

    d. How is the fraction related to the number of squares in a group?

5. How can you find the number of different possible subsets (combinations) of *m* items that can be selected from a set of *n* items?

Chapter 7 Summary

Probabilities are used to help interpret and understand situations involving uncertainty. There are two kinds of probabilities: experimental and theoretical. *Experimental probabilities* are determined by observing the outcome of a situation or experiment over a large number of trials. For example, meteorologists collect weather data over an extended period of time. They may use this data to predict the amount of rainfall expected during a given month. Information like this can be very helpful when dealing with potential water shortages.

Theoretical probabilities are based on mathematical analysis of the possible outcomes of an experiment rather than on observation of the outcomes. A fundamental concept of probability is that over a large number of trials, the experimental probability of an event will approach the theoretical probability. This idea was emphasized throughout the activities by having you compare your individual results from an experiment with the theoretical results and then combining your results with other people's to obtain the results for a larger number of trials and repeating the comparison. In general, as the number of trials increased, the experimental probability more closely approximates the theoretical probability.

The concept of probability was introduced in Activity 1 by building on intuitive ideas about chance. Activity 2 introduced the idea of a *fair game*. The intuitive notion is that a game is fair if the chances of winning are equal to the chances of losing, or if all players in a game have the same chance of winning. The notion of a fair game was explored further in Activity 5 where games were analyzed using matrices and probability trees, tools that are useful in determining probabilities.

Many of the basic concepts of probability were introduced in Activity 3 and followed up in Activity 4. You learned that a probability is a ratio that expresses the likelihood of something happening. You also explored the distinction between experimental and theoretical probabilities. Some of the basic ideas introduced in Activity 3 are that

$$\text{Experimental probability} = \frac{\text{number of times an event occurs}}{\text{total number of trials}}$$

and that if the outcomes of an experiment are equally likely, then

$$\text{Theoretical probability} = \frac{\text{number of outcomes making up the event}}{\text{total number of possible outcomes}}.$$

If an event **cannot happen**, its probability is zero, and if an event **is certain to happen**, its probability is one. If A and B are *complementary events*, then $P(A) = 1 - P(B)$. And finally, if C and D are *mutually exclusive events*, that is, events that cannot occur simultaneously, then $P(C \text{ or } D) = P(C) + P(D)$.

Activities 4 and 9 explored the concepts of *expected value* and *random sampling*. If you know the probability that an event will occur, then you can estimate how many times it will occur in a given number of trials—the expected value—by multiplying the number of trials by the probability. Thus, if we know the probability that a nineteen-year-old woman has never been married is 0.887, we would estimate that in a group of 25 nineteen-year-old women, about 22 (0.887×25) will never have been married.

In Activity 4, the probabilities that a bag contains particular colors of squares were estimated by *random sampling*, that is, drawing squares from the bag and recording the outcomes. The probabilities were then used to predict how many squares of each color were in the bag. Random sampling was also used in Activity 9 to determine how many faces of a die have particular numbers on them. The idea of predicting the characteristics of a population (in these cases, the squares in a bag and the six faces of a die) based on a random sample will be explored again in Chapter 8.

Another topic related to sampling is the distribution of outcomes. In Activity 6, you investigated how the distribution of the outcomes of an experiment is affected by the probabilities of the outcomes. The distribution is often a bell-shaped curve centered on the outcome or outcomes with the highest probabilities. The curve can be skewed to the left or right depending on which outcome has the greatest probability.

In Activity 5, a tree diagram was used to determine the probabilities of the outcomes of a multi-stage experiment. In the activity, you learned that the probability of each outcome is the product of the probabilities along the path leading to it. Activity 7 showed how the probabilities of certain events can be determined using Pascal's Triangle.

Many real-world probability problems are too complex to analyze theoretically and too difficult, time consuming, or expensive to observe actual trials. In these cases, the solution can often be found by simulating the problem. Activity 8 developed the techniques used to design and conduct a *simulation*. Because a large number of trials must be conducted to obtain accurate results by simulating an experiment, computers are usually used to carry out simulations.

Finally, the use of permutations and combinations to count the outcomes of an experiment was introduced in Activity 10.

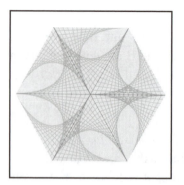

Chapter 8
Statistics: An Introduction

"The amount of data available to help make decisions in business, politics, research, and everyday life is staggering: Consumer surveys guide the development and marketing of products. Polls help determine political campaign strategies, and experiments are used to evaluate the safety and efficacy of new medical treatments. Statistics are often misused to sway public opinion on issues or to misrepresent the quality and effectiveness of commercial products. Students need to know about data analysis and related aspects of probability in order to reason statistically—skills necessary to becoming informed citizens and intelligent consumers."

—Principles and Standards for School Mathematics

Statistics may be defined as the science of collecting, organizing, and interpreting data. In this chapter, you will learn to collect and sample data and to organize data using a variety of techniques—line plots, frequency tables, bar graphs, stem-and-leaf plots, and box-and-whisker plots. You also will learn to describe data using measures of central tendency and to interpret data presented in graphs.

The activities in this chapter are designed to develop your understanding of these concepts and the processes of statistics. The emphasis in the activities is on the visual presentation of data and informal methods of data analysis rather than on formal statistical methods. In the activities, you will learn to use statistics to communicate information about sets of data effectively. You will also learn to interpret statistical displays and to make critical and informed decisions based on them.

Activity 1: Graphing *m&m's*®

PURPOSE	Use a variety of graphs to display data and explore relationships among the data.
MATERIALS	A one-pound bag of *m&m's* Plain Chocolate Candies*, a one-tablespoon measure, colored pencils, a balance scale and weights, six line plot charts (one for each color of *m&m's*, page A-38), and a calculator
GROUPING	Work individually and as a whole class.
GETTING STARTED	*m&m's*® Plain Chocolate Candies come in six colors: brown, green, orange, red, blue, and yellow.

Before you take a sample from the bag of *m&m's*, answer the following.

1. a. Which color of *m&m's* do you think will occur most often in the bag? Why?

    b. in your sample? Why?

2. a. Which color of *m&m's* do you think will occur least often in the bag? Why?

    b. in your sample? Why?

REAL GRAPHS

1. Take a sample of *m&m's* by dipping the measuring spoon into the bag of candy and removing a spoonful. *CAUTION: Do not eat any of the m&m's!*

Statistical data are often displayed graphically. Using a graph rather than simply presenting the data as a set of numbers makes it easier to study relationships in the data.

2. Arrange the *m&m's* in your sample on Graph 1 on the next page. This type of graph is often called a *real graph* because the statistical data are displayed using the actual objects whose frequencies are being compared.

3. Record the number of *m&m's* of each color and the total number in your sample.

    Brown: ____ Orange: ____ Blue: ____ Green: ____ Red: ____ Yellow: ____ Total: ____

4. What color occurred most often and what color occurred least often in your sample? How do these colors compare with your predictions?

**m&m's* Plain Chocolate Candies is a registered trademark of Mars, Inc.

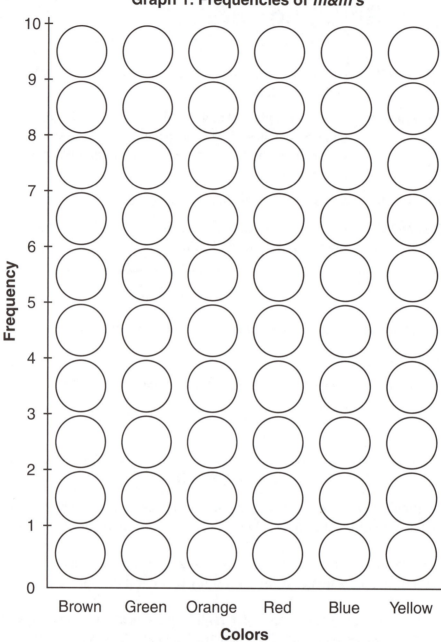

Graph 1: Frequencies of *m&m's*

PICTOGRAPHS

1. As you remove each candy from the graph, color its circle the appropriate color. This type of graph is called a *pictograph* because the data are displayed using parallel columns (or rows) of pictures in which each picture represents one or more of the objects being compared.

Now you may eat the m&m's in your sample!

2. Compare your pictograph with your classmates' pictographs. Describe any similarities and differences and explain why these may have occurred.

LINE PLOTS

1. Complete the following to collect the class data for the yellow *m&m's*.

   a. What is the maximum number of yellow *m&m's* in anyone's sample?

   b. What is the minimum number of yellow *m&m's* in anyone's sample?

   c. Title one of the line-plot charts "Yellow" and use the maximum and minimum values from Parts (a) and (b) to label the scale on the number line.

   d. Each time a person reports the number of yellow *m&m's* in his or her sample, record an X above that number on the number line.

   Example: **Yellow *m&m's***

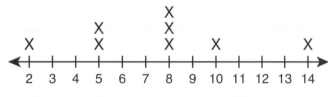

   Minimum = 2   *Maximum = 14*

This type of graph is called a *line plot*. Line plots provide a quick, simple way to organize numerical data. They work best when there are fewer than 25 data points.

2. Repeat Exercise 1 for each color of *m&m's*.

3. Use the line plots to describe the data for each color. Rather than just looking at individual numbers, describe the shape of the data—any patterns or special features such as clusters or gaps in the data and isolated data points—that tell how the data are distributed.

4. Use the line plots to find the total number of each color of *m&m's* in the samples.

   Brown: _____ Orange: _____ Blue: _____ Green: _____ Red: _____ Yellow: _____

PREDICTIONS

1. Use the class data to predict the number of each color of *m&m's* that you would expect to find in a one-pound bag.

   Brown: _____ Orange: _____ Blue: _____ Green: _____ Red: _____ Yellow: _____

2. Describe the procedure you used to make your predictions.

3. Help your classmates count the *m&m's* remaining in the bag. Add these counts to the numbers you already have. What was the total number of each color of *m&m's* in the bag?

   Brown: _____ Orange: _____ Blue: _____ Green: _____ Red: _____ Yellow: _____

4. How do these totals compare with the predictions you made in Exercise 1?

PICTOGRAPHS REVISITED

Construct a pictograph for the number of each color of *m&m's* in the bag on Graph 2 below. **HINT:** Let each circle represent more than one *m&m*.

Graph 2: Frequencies of *m&m's*

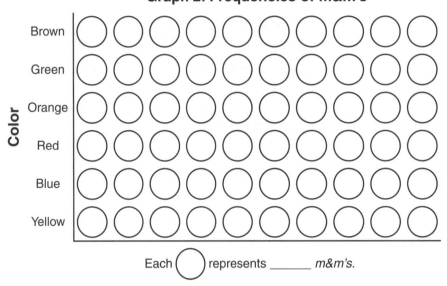

Each ◯ represents _____ *m&m's*.

BAR GRAPHS

1. Use Graph 3 below to construct a horizontal bar graph for the number of each color of *m&m's* in the bag. Label the scale on the horizontal axis.

Graph 3: Frequencies of *m&m's*

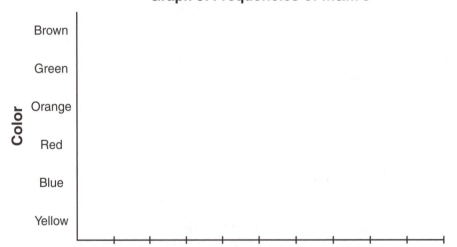

2. a. Which graph, the pictograph or the bar graph, was easier to construct? Why?

   b. Which graph is easier to read? Why?

Activity 2: What's in the Bag?

PURPOSE   Use percents to compare the composition of a sample and the composition of the population.

MATERIALS   Data from Activity 1 on the number of each color of *m&m's* Plain Chocolate Candies in a one-pound bag

GROUPING   Work individually.

GETTING STARTED

> ### DID YOU KNOW?
> According to Mars, Inc., the manufacturers of *m&m's* Plain Chocolate Candies, there are 30% brown, 20% yellow, 20% red, 10% orange, 10% blue, and 10% green candies in each bag.

1. Record the number of each color of *m&m's* in the one-pound bag from Activity 1 in the table below. Determine the total number of *m&m's* and the percent of each color.

Color	Brown	Green	Orange	Red	Blue	Yellow	Total
Number							
Percent							

2. Use the data in the table to support or dispute the claim made by the manufacturer. How could the company explain any discrepancies between its claim and the percents you found?

EXTENSIONS   Investigate the effect of using different sampling techniques by repeating the experiment in Activity 1 using a one-pound bag of individually wrapped FUN SIZE packs of *m&m's* and giving each student a FUN SIZE pack for his or her individual sample.* Answer the following questions.

1. Are the percents of the different colors in your FUN SIZE pack the same as the percents you found in your individual sample in Activity 1?

2. In which individual sample (your scoop of *m&m's* or the FUN SIZE pack) are the percents of the colors closest to the percents given by the manufacturer?

3. Are the percents of the colors in the one-pound bag of FUN SIZE packs the same as the percents you found in the one-pound bag of *m&m's* in Activity 1? Are they the same as the percents given by the manufacturer?

*FUN SIZE® is a registered trademark of Mars, Inc.

Activity 3: Gaps and Clusters

PURPOSE	Interpret data by identifying gaps and clusters.
GROUPING	Work individually.
GETTING STARTED	The average life spans of several selected mammals are shown in the following table.

Average Life Spans of Selected Mammals

Animal	Life Span (yrs)	Animal	Life Span (yrs)
Bear	15–30	Hippopotamus	30
Cat	10–12	Horse	20–25
Cow	9–12	Lion	10
Deer	10–15	Monkey	12–15
Dog	10–12	Mouse	1–3
Elephant	30–40	Pig	10
Fox	8–10	Rat	3
Goat	12	Sheep	12
Guinea Pig	3	Squirrel	8–9
Hamster, golden	2	Wolf	10–12

SOURCE: 1996 Information Please Almanac.

1. Make a line plot of the life spans by plotting either the given value or the midpoint of the range of values for each animal.

Use your line plot to answer the following questions.

2. A *gap* in a set of data is a large interval that doesn't contain any data points. One gap occurs between 3 and 8.5 years. Where else do you see gaps in the data?

3. a. A *cluster* is an isolated group of data points. The life spans from 8.5 to 13.5 years form a cluster. In comparison to the other animals in the table, how would you describe the size of the animals that have life spans from 8.5 to 13.5 years?

   b. Where else do you see clusters in the data? How would you describe the animals with life spans in those clusters?

4. Use the table and your line plot to help estimate the average life span of each of each animal below. Explain your answers.

   a. antelope   b. giraffe   c. chipmunk

Activity 4: Grouped Data

PURPOSE	Display data using grouped frequency tables and stem-and-leaf plots.
GROUPING	Work individually or in pairs.
GETTING STARTED	In many situations, there may be so much data, or the data may be so spread out, that it becomes difficult to construct a line plot or a frequency table for individual items. In such cases, it may be more convenient to group the data.

President	Political Party[1]	Age at Inauguration	President	Political Party[1]	Age at Inauguration
Washington	F	57	Cleveland	D	47
J. Adams	F	61	B. Harrison	R	55
Jefferson	DR	57	Cleveland	D	55
Madison	DR	57	McKinley	R	54
Monroe	DR	58	T. Roosevelt	R	42
J. Q. Adams	DR	57	Taft	R	51
Jackson	D	61	Wilson	D	56
Van Buren	D	54	Harding	R	55
W. H. Harrison	W	68	Coolidge	R	51
Tyler	W	51	Hoover	R	54
Polk	D	49	F. D. Roosevelt	D	51
Taylor	W	64	Truman	D	60
Fillmore	W	50	Eisenhower	R	62
Pierce	D	48	Kennedy	D	43
Buchanan	D	65	L. B. Johnson	D	55
Lincoln	R	52	Nixon	R	56
A. Johnson	U	56	Ford	R	61
Grant	R	46	Carter	D	52
Hayes	R	54	Reagan	R	69
Garfield	R	49	Bush	R	64
Arthur	R	51	Clinton	D	46

[1]F = Federalist, DR = Democratic-Republican, D = Democrat, W = Whig, R = Republican, U = Unionist
SOURCE: The World Almanac and Book of Facts 1992.

GROUPED FREQUENCY TABLES

Complete the *grouped frequency table* for the presidents' ages at inauguration.

How many different ages can fall within an interval in the table?

Can you determine the range, median, and mode of the ages from the information in the table? Explain why or why not for each statistic.

Age at Inauguration		
Interval	Tally	Frequency
40–44		
45–49		
50–54		
55–59		
60–64		
65–69		

STEM-AND-LEAF PLOTS

Stem-and-leaf plots provide a natural way to group data in intervals. To construct a stem-and-leaf plot, first find the smallest and the largest data points.

What is the youngest age at inauguration for the presidents? _____ the oldest age? _____

Next, decide on the stems. Since the presidents were inaugurated in their 40s, 50s, and 60s, use the tens digits of the ages as the stems. Write the stem digits in a column from least to greatest on the left side of a vertical line, as shown.

Stem	Leaf
4	
5	
6	

For each age at inauguration, record a leaf by writing the units digit of the age on the right side of the vertical line in the row that contains its stem. The leaves for Washington and J. Adams have been done for you.

Stem	Leaf
4	
5	7
6	1

After all the leaves have been recorded, rearrange the leaves in increasing order.

Use the stem-and-leaf plot to find the range, median, and mode of the ages at inauguration.

How are a stem-and-leaf plot and a grouped frequency table alike? How do they differ?

Stem	Leaf
4	
5	
6	

On the grid, list the stems for the age at inauguration in the center column. Construct a stem-and-leaf plot for the age at inauguration of the Democratic (D) presidents to the left of the stems and one for the Republican (R) presidents to the right of the stems.

How do the ages at inauguration of Democratic presidents compare to those of Republican presidents?

Democratic		Republican

Activity 5: What's the Average?

PURPOSE Investigate mean, median, and mode and examine how each average is affected by extremes in the data.

MATERIALS 20 strips of one-centimeter graph paper (page A-49) and a pair of scissors for each group

GROUPING Work individually or in groups of 3 or 4.

GETTING STARTED When we describe a set of data, it is often convenient to use a single number, often called the *average*, to indicate where the data are centered or concentrated. The mean, the median, and the mode are three commonly used *averages*.

Write the name of each of the following states on a strip of graph paper. Use one strip of paper for each state and one square for each letter in the name. Cut off the unused squares on the end of each strip.

Arizona, Hawaii, Ohio, Maine, Oregon, Idaho, Texas, Louisiana, Kentucky

Arrange the names from shortest to longest, as shown in the example.

Count the number of letters in the name of each state. On a separate strip of graph paper, **write the numbers in order from least to greatest.** Write one number in each square and do not leave any blank squares between numbers. Cut off the unused squares on the end of the strip.

Example:

N	E	V	A	D	A
M	O	N	T	A	N	A
V	I	R	G	I	N	I	A		6	7	8	9	9
W	I	S	C	O	N	S	I	N
M	I	N	N	E	S	O	T	A

THE MODE

Look at the numbers on the strip. What number of letters occurs most often in the names of the states?

This is the **mode** of the numbers of letters in the names of the states.

In the example, the mode is 9.

N	E	V	A	D	A
M	O	N	T	A	N	A
V	I	R	G	I	N	I	A

Mode (9 letters)
| W | I | S | C | O | N | S | I | N |
| M | I | N | N | E | S | O | T | A |

THE MEDIAN

1. Fold the strip containing the numbers of letters in the names of the states in half by folding the ends together.

2. Unfold the strip. Through which number does the fold pass?

This is the **median** number of letters in the names of the states. In the example, the median is 8.

3. If the fold is on the line between two numbers, what number would you use for the median? Why?

4. How many states have names that contain fewer letters than the median? more letters than the median?

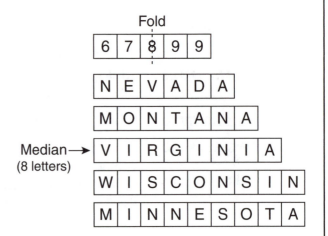

Fold

| 6 | 7 | 8 | 9 | 9 |

| N | E | V | A | D | A |

| M | O | N | T | A | N | A |

Median→ (8 letters) | V | I | R | G | I | N | I | A |

| W | I | S | C | O | N | S | I | N |

| M | I | N | N | E | S | O | T | A |

THE MEAN

To find the **mean** of the numbers of letters in the names, cut off letters from the longer names and move them to fill in the shorter ones. Continue cutting off and moving letters until all the rows have as close to the same number of letters as possible.

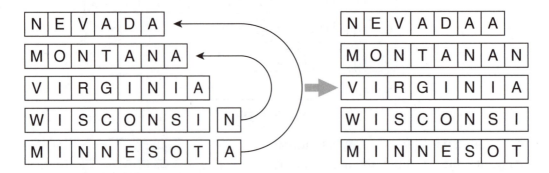

The mean in this example is a little less than eight because all the rows except one contain eight letters.

What is the mean of the numbers of letters in the names of the nine states?

1. Write each letter of Massachusetts in a square on a strip of graph paper. Add this to the data for the other nine states. Then repeat the steps to find the median, mode, and mean of the numbers of letters in the names of the ten states.

   The median is _____.   The mode is _____.   The mean is _____.

2. Compare these averages with those for the original nine states. Describe how the addition of Massachusetts affected each average and explain the differences.

3. Remove the data for Massachusetts. Write the names Maryland, Michigan, and Oklahoma on strips of graph paper. Add them to the data for the original nine states. Then repeat the steps to find the median, mode, and mean.

   The median is _____.   The mode is _____.   The mean is _____.

4. Compare these averages with those for the original nine states. Describe how these additions affected each average and explain the differences.

5. To find the mean of the numbers of letters in the names of *N* states, the letters making up the names of the states must be separated into *N* sets with the same (or nearly the same) number of letters in each set. Explain how you could find the number of letters in each set without writing each letter on a square.

WHICH WOULD YOU USE?

Sam Slugger's contract with the Columbus Mudcats baseball team says his annual salary will be $1,000,000 times the ***average*** of his batting averages for the preceding five seasons. Sam's batting averages for the past five seasons were .145, .130, .160, .130, and .495.

1. If you were Sam, which *average*—mean, median, or mode—would you want to use to compute your salary? Why?

2. If you were the owner of the Mudcats, which *average* would you want to use? Why?

3. Sam's contract went to arbitration. You are the arbitrator. Which *average* would you use to determine Sam's salary? How would you justify your decision?

IDENTIFY THE AVERAGE

Which *average*—mean, median, or mode—do you think was used in each of the following statements? Explain your choice in each case.

The *average* lady's shoe size is $7\frac{1}{2}$.

The *average* size of a household in the United States is 2.64 people.

The *average* annual family income in the United States is $40,611.

Activity 6: Finger-Snapping Time

PURPOSE	Display and compare data using box-and-whisker plots.
MATERIALS	Scissors, three strips of one-centimeter graph paper (page A-49), and a clock or watch to measure elapsed time in seconds
GROUPING	Work in pairs.
GETTING STARTED	Snap your fingers as fast as you can for 15 seconds. Have your partner time you while you snap your fingers and count the number of snaps. Then do the same thing for your partner. Record the data in the table below.

Ask 13 other people how many times they snapped their fingers in 15 seconds and record the information in the table.

Finger Snaps I

Person	Finger Snaps in 15 sec	Person	Finger Snaps in 15 sec	Person	Finger Snaps in 15 sec
You		4		9	
Partner		5		10	
1		6		11	
2		7		12	
3		8		13	

Box-and-whisker plots provide a useful method for summarizing and comparing data such as the number of times people can snap their fingers.

- **Step 1 in constructing a box-and-whisker plot is to order the data values from least to greatest.**

  1. Write the finger-snapping data in order from least to greatest on a strip of graph paper. Write one number in each square. Do not leave any blank squares between numbers. Cut off the unused squares on the end of the strip.

Example:

20	35	37	39	41	45	45	47	59

LE = 20; UE = 59

- **Step 2—find the extremes of the data.** The smallest data point is called the *lower extreme* (LE), and the largest data point is called the *upper extreme* (UE).

  2. Find and record the lower extreme and the upper extreme of the finger-snapping data.

  LE = _____   UE = _____

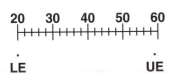

LE   UE

3. Use the extremes to select an appropriate scale and label the number line below.

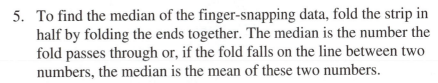

4. Locate the lower and upper extremes by marking a dot under their coordinates on the scale, as in the example.

- **Step 3—find the median of the data.**

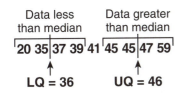

Median = 41

5. To find the median of the finger-snapping data, fold the strip in half by folding the ends together. The median is the number the fold passes through or, if the fold falls on the line between two numbers, the median is the mean of these two numbers.

6. Record the median and mark its location by making a dot under its coordinate on the scale.

   Median = _____

- **Step 4—find the quartiles for the data.**

The *lower quartile* (LQ) is the median of the data values that are less than the median.

Data less   Data greater
than median   than median

20 35 | 37 39 | 41 | 45 45 | 47 59

LQ = 36   UQ = 46

7. Look at the part of the strip that contains data values that are less than the median. Find the median of these values by folding this part of the strip in half. This is the lower quartile of the data.

   Lower Quartile = _____

The *upper quartile* (UQ) is the median of the data values that are greater than the median of the data.

8. Look at the part of the strip that contains data values that are greater than the median. Find the median of these values by folding this part of the strip in half. This is the upper quartile of the data.

   Upper Quartile = _____

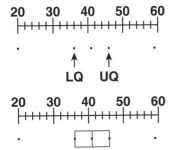

LQ   UQ

9. Mark the locations of the upper and lower quartiles by making a dot for each below its coordinate on the scale.

10. Form a box by drawing vertical segments through the dots for the upper and lower quartiles and connecting the endpoints of the segments, as in the example. Then draw a vertical segment through the median as shown.

• **The final step is to identify and plot any outliers in the data.**

IQR = 46 − 36 = 10

11. Find the difference between the upper and lower quartiles. This difference is known as the *interquartile range* (IQR).

Interquartile Range = UQ − LQ = _____

If a data point is more than 1.5 interquartile ranges greater than the upper quartile or more than 1.5 interquartile ranges less than the lower quartile, it is called an *outlier*.

In the example, LQ − 1.5 × IQR = 36 − 1.5 × 10 = 36 − 15 = 21.

Thus 20 is an outlier, since 20 < LQ − 1.5 × IQR.

12. Identify any outliers in your data. Mark their locations by making a dot for each one below its coordinate on the scale.

13. Complete the plot by drawing segments from the least data point that is not an outlier to the lower quartile and from the upper quartile to the greatest data point that is not an outlier. These segments are the whiskers.

14. Study the completed box-and-whisker plot. About what percent of the data points lie between the

   a. lower extreme and the lower quartile?

   b. upper extreme and the upper quartile?

   c. lower quartile and the median?

   d. median and the upper quartile?

15. About what percent of the data points lie in the box?

16. Repeat the finger-snapping experiment, only this time snap your fingers as fast as you can for 30 seconds. Divide the number of snaps by 2 to get your rate per 15 seconds Record your rate, your partner's rate, and the rates for 13 other people in the table below.

Finger Snaps II

Person	Sex	Finger Snaps per 15 sec	Person	Sex	Finger Snaps per 15 sec	Person	Sex	Finger Snaps per 15 sec
You			4			9		
Partner			5			10		
1			6			11		
2			7			12		
3			8			13		

17. a. Construct a box-and-whisker plot for the data in the table Finger Snaps II.

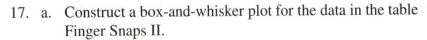

   b. In the space below the plot constructed in 17a, re-construct a box-and-whisker plot using the data for Finger Snaps I, page 174, and the scale in 17a.

18. a. How do the medians of the two sets of data compare?

   b. The extremes?

   c. The interquartile ranges?

   d. The upper quartiles?

   e. The lower quartiles?

19. Based on your observations in Exercise 18, how do the finger-snapping rates in the first experiment compare to the rates in the second experiment? How would you explain any similarities or differences?

EXTENSIONS

1. a. Separate the data in the Finger Snaps II table into rates for males and rates for females. Construct box-and-whisker plots for the male and female rates using the same scale for both.

   b. Based on your graphs, do you think there is any difference between the finger-snapping rates for males and for females? Explain your answers.

2. a. Construct a stem-and-leaf plot for the data for Finger Snaps I.

   b. What can you learn about the distribution of the data (gaps, clusters, extremes, outliers, etc.) and the averages **from both** the box-and-whisker plot and the stem-and-leaf plot of the data?

   c. What information can you get about the distribution of the data and the averages from a box-and-whisker plot **but not from** a stem-and-leaf plot?

   d. What information can you get about the distribution of the data and the averages from a stem-and-leaf plot **but not from** a box-and-whisker plot?

Activity 7: The Weather Report

PURPOSE   Apply the concepts of data analysis, evaluate statements based on data, and analyze the advantages and disadvantages of using different displays of data.

MATERIALS   A calculator and graph paper

GROUPING   Work individually or in pairs.

Normal Daily Temperature for Some U.S. Cities

Based on the Period 1950–1980

	Jan	Feb	Mar	Apr	May	Jun	July	Aug	Sep	Oct	Nov	Dec
Los Angeles, CA	56	57	57	60	62	66	69	70	70	66	61	57
San Francisco, CA	49	52	53	55	58	61	62	63	64	61	55	49
Wichita, KS	30	35	44	56	66	76	81	80	71	59	44	34
Portland, ME	22	23	32	43	53	62	68	67	59	48	38	26
Portland, OR	39	43	46	50	57	63	68	67	63	54	46	41
Charlotte, NC	41	43	50	60	68	75	79	78	72	61	51	43
Seattle, WA	39	43	44	49	55	60	65	64	60	53	45	41
Spokane, WA	26	32	38	46	54	62	70	69	59	48	35	29

SOURCE: U.S. National Oceanic and Atmospheric Administration.

1. Construct a stem-and-leaf plot for the weather data given for San Francisco and Wichita.

San Francisco	Wichita
4	2
5	3
6	4
	5
	6
	7
	8

2. Compute the mean, median, and mode for the data given for San Francisco and Wichita.

	San Francisco	Wichita
Mean		
Median		
Mode		

3. Construct line graphs for the temperatures for San Francisco and Wichita on the grid below.

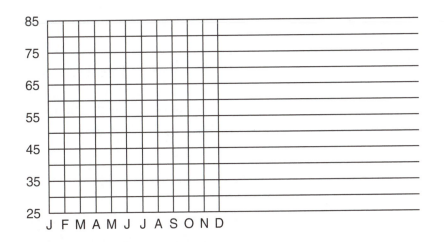

4. Using the scale on the grid above, construct vertical box-and-whisker plots for the temperatures for San Francisco and Wichita to the right of the line graphs.

5. Draw a horizontal line through the graphs to show the median for each set of data.

6. For how many months of the year is the temperature in Wichita within the range of the temperatures in San Francisco?

7. Which display of data can most easily be used to answer Exercise 6? Explain why or why not for each display.

8. The Wichita Chamber of Commerce could advertise that "the average annual temperature in Wichita is the same as that in *balmy* San Francisco." Evaluate this claim on the basis of the actual temperature data.

9. The mean and median annual temperatures for these two cities are nearly the same. Compare how accurately the mean and the median describe the climate in each city.

10. What other information do you need in addition to the medians or means to accurately describe the annual temperatures?

11. Find the latitude of Wichita: ____; and of San Francisco: ____. Which city is farther north?

12. If the difference in latitude is not significant, explain how the geographical location of each city affects its annual temperature.

13. Construct a back-to-back stem-and-leaf plot for Portland, ME, and Portland, OR.

Portland, ME		Portland, OR
	2	
	3	
	4	
	5	
	6	
	7	
	8	

14. On the grid below, construct line graphs and accompanying vertical box-and-whisker plots for the weather data for the two cities. Then draw the median lines. Explore questions similar to Exercises 6–12 for these two sets of data.

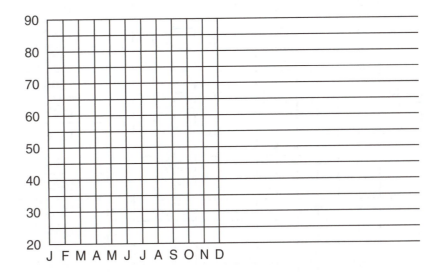

15. Choose other pairs of cities from the table and construct various displays for the weather data. Explore similarities and differences in the data. Determine the latitude and geographical location of each city. Explain how these affect the similarities or differences in the climates of the chosen cities.

Activity 8: Are Women Catching Up?

PURPOSE   Display and analyze data using line graphs, scatter plots, and median fit lines. Make conjectures based on data and defend the conjectures.

GROUPING   Work individually or in pairs.

GETTING STARTED   In 1992, researchers at the University of California, Los Angeles, School of Medicine published an analysis of data indicating that within 65 years, top female and male runners might perform equally well in the 200-m, 400-m, 800-m, and 1500-m runs. Other researchers disputed the claim. What do you think? Let's look at some of the data.

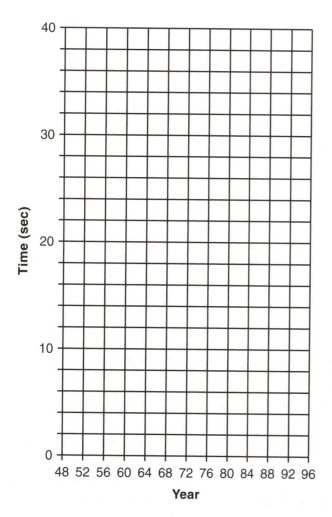

Olympic Track Records: 200-m Run

Year	Men's Time (sec)	Women's Time (sec)
1948	21.1	24.4
1952	20.7	23.7
1956	20.6	23.4
1960	20.5	24.0
1964	20.3	23.0
1968	19.83	22.5
1972	20.00	22.40
1976	20.23	22.37
1980	20.19	22.03
1984	19.80	21.81
1988	19.75	21.34
1992	20.01	21.81
1996	19.32	22.12

SOURCE: The New York Times ALMANAC 1999.

1. Draw line graphs for the men's and the women's record times for the 200-m run on the grid at the left.

2. Based on the graphs, do you agree with the conclusion of the UCLA researchers about the men's and women's record times for the 200-m run? Why or why not?

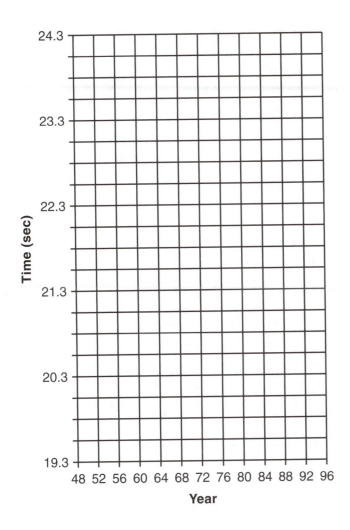

3. Redraw the line graphs for the men's and the women's record times for the 200-m run on the grid at the left.

4. Based on these graphs, what might you conclude about the men's and women's record times for the 200-m run? Why?

5. Is your conclusion in Exercise 4 the same as that in Exercise 2? Explain.

6. Both pairs of graphs present the same data on the same-sized grid. Why do they look so different?

When data appear to lie roughly along a straight line, it is often possible to fit a line to the data and use the line to make predictions. The following exercises develop one technique for doing this. *Note:* You must always be cautious about fitting a line to data, because the data may actually fall near a curve or be clustered in two or more regions.

7. The grid on page 184 contains a scatter plot of the women's record times for the 200-m run. Draw two vertical lines that divide the data into three groups with approximately the same number of data points in each group. If the data cannot be divided evenly, the two outer groups should contain the same number of data points.

8. The group on the left contains four data points: (1948, 24.4), (1952, 23.7), (1956, 23.4), and (1960, 24.0). The median of the years is 1954, and the median of the times is 23.85 sec. Use a plus symbol (+) to mark the point (54, 23.85) on the grid.

9. Find the median of the years and the median of the times for the data points in each of the other two groups, as in Exercise 8. Use a plus symbol (+) to mark the corresponding points on the grid.

10. Place a ruler so that it passes through the two plus marks (+) in the outside groups. Then, keeping the ruler parallel to the line through the two outer plus marks (+), slide it one-third of the way to the middle symbol (+) and draw a line. This is the *median fit line* for the data.

11. Make a scatter plot for the men's times on the grid. Repeat Exercises 7–10 to construct the median fit line for this data.

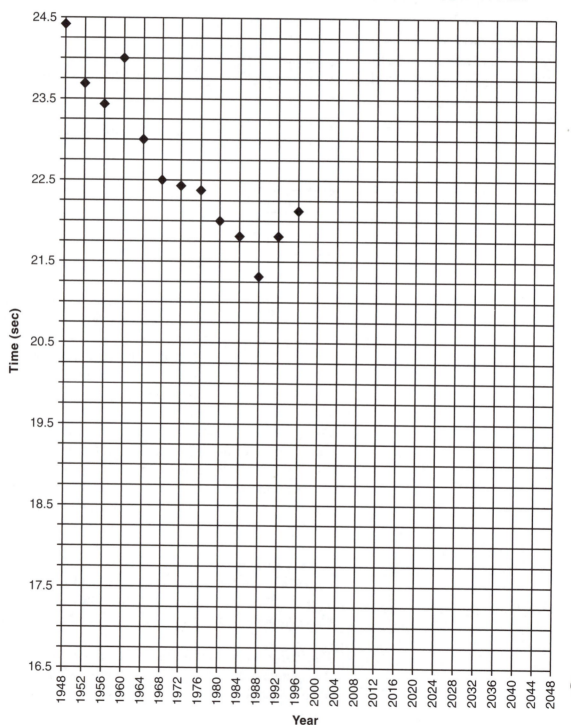

12. a. Use the median fit lines to predict the year when women will have the same time in the 200-m run as men.

    b. How does your prediction in Part a compare with that made by the University of California researchers?

    c. Approximately what will be the record for the 200-m run when men's and women's times are equal?

13. Why are women's times for the 200-m run decreasing at a faster rate than men's times? Explain.

14. Prepare an argument that supports the prediction that in the future, women's times for the 200-m run will equal men's times.

15. Prepare an argument to support the position that women's times for the 200-m run will never equal men's times?

EXTENSIONS   Find the men's and women's winning times for the Boston Marathon over the last 20 years.

1. Use line graphs to construct two displays of the data, one that could be used to argue that in the future women's times for the marathon will be equal to the men's, and one that could be used to argue that they will not.

2. Construct scatter plots of the data. Find the median fit lines for both sets of data. Use the lines to predict the year that the women's time for the Boston Marathon will equal the men's.

3. Compare the data using box-and-whisker plots. Explain how the two different visual presentations of the data can lead to different conclusions.

Activity 9: To Change or Not to Change

PURPOSE   Analyze a set of data and prepare a proposal based on the results.

GROUPING   Work individually.

GETTING STARTED   There has been a great deal of debate in recent years about the need to lengthen the school day or to extend the school year for students in the United States. What's your opinion?

The following table contains data on the average mathematics scores on the International Assessment of Educational Progress for 13 year-old students in various countries. It also includes information on the length of the school year and the length of the school day in each country.

Country	Average Percent Correct	Average Days of Instruction per Year	Average Minutes of Instruction per School Day
Canada	62	188	304
Emilia-Romagna, Italy	64	204	289
France	64	174	370
Hungary	68	177	223
Ireland	61	173	323
Israel	63	215	278
Jordan	40	191	260
Korea	73	222	264
Scotland	61	191	324
Slovenia	57	190	248
Soviet Union	70	198	243
Spain	55	188	285
Switzerland	71	207	305
Taiwan	73	222	318
United States	55	178	338

Source: *Educational Testing Service,* Learning Mathematics, *Feb. 1992.*

Use the statistical concepts you have studied to organize and analyze the data in the table. Based on the results, prepare a report recommending what changes, if any, should be made in the length of the school year and the school day for students in the United States. Whenever appropriate, use averages and statistical displays to support and clarify your position.

Chapter 8 Summary

Activities 1, 4, and 6 introduced a variety of graphs, tables, and plots that can be used to organize and display data. The focus of the activities was on the potential uses and limitations of each statistical display, not just the construction of each display.

In general, *real graphs* do not provide a practical method for displaying statistical data. *Pictographs* are common in the popular media, probably because of their aesthetic appeal, but they are difficult to interpret and construct. *Bar graphs*, on the other hand, display exactly the same information and have the advantage of being easy to construct and to interpret.

Because *line plots* are quick and easy to construct, they provide a very useful tool for preliminary data analysis. Maximum and minimum values, clusters and gaps in the data, outliers, the median and the mode are all easily identified on a line plot. The major limitation is that line plots are convenient to use only with relatively small sets of data.

Stem-and-leaf plots provide information about how data are distributed. Maximum and minimum values, the median and mode, clusters and gaps in the data, and outliers are all easily identified in a stem-and-leaf plot. *Box-and-whisker plots* focus on the range and distribution of the data. Clustering of the data can sometimes be identified, but gaps in the data cannot. Box-and-whisker plots are very useful when working with large sets of data and for comparing different sets of data.

Activity 5 introduced the concept of an average and developed the meanings of the mean, median, and mode. The development focused on (a) how each measure provides an indication of where data are centered or concentrated, (b) how each measure is affected by extremes, and (c) how the physical modeling is related to the algorithm for computing the mean.

The activity illustrated that the *mean* has a leveling or smoothing effect. Another way of expressing this is, if all the data had the same value, the data points would all equal the mean. The mean is the most commonly used average—in fact, many people erroneously use the terms *average* and *mean* synonymously. However, the value of the mean may be affected by extremes in the data, and therefore it may not be the most representative value to use for an average.

The *median* is the middle data point or the mean of the two middle data points. The value of the median is not significantly affected by unusually large or small data points; thus the median is a more appropriate average than the mean when there are extremes in the data. However, the median may not accurately reflect concentrations in the data.

The *mode*, the data point that occurs most often, is a measure of where the data are concentrated. Its usefulness is limited, however, since the frequency of occurrence of the mode may not be significantly different than that of other data points, the mode may be an outlier, or the data may have more than one mode.

Activities 1 and 2 explored the relationship between a sample and the population. Predicting the characteristics of a population from a sample is an important concept in statistics. The reliability of such predictions is affected by two factors: the sample size and the randomness of the sample. Assuming the samples are selected randomly, predictions generally become more accurate as the size of the sample increases. Randomness was demonstrated in Activity 1 by the fact that there was a great deal of variation in the individual samples of *m&m's*.

Identifying and interpreting gaps and clusters in data was explored in Activity 3. The focus of Activity 7 was on analyzing various plots to determine what information can be more easily derived from one display than another and to evaluate the use of mean or median as a good descriptor of average.

Advertisers, policy makers, and decision-makers constantly use the term *average*, but the average reported may not accurately describe the entire set of data. It is very important to know the range of data to fully understand what the mean or median really indicates. For example, the mean and median temperatures for San Francisco and Wichita differ by only a degree. However, the range of the data varies considerably.

The stem-and-leaf plots illustrate the difference in the range of data, but the difference is much more visually apparent in line graphs or in side by side box-and-whisker plots. The median and the upper and lower quartiles of data can be identified in a stem-and-leaf plot; however, the relationships among these measures are much clearer in box-and-whisker plots. Each display has its own unique characteristics, and each offers a different insight into the data.

Since each display provides a different insight, the way data is displayed can affect the conclusions drawn from it. As illustrated in Activity 8, the conclusions can also be influenced by how the display is constructed. The choice of scales for a line graph, for example, can dramatically alter the visual impact of the data and result in two different interpretations. Activity 9 provided an opportunity for you to apply statistical concepts in a real world context to display and interpret data, to reason from the data, to make a decision based on your interpretation, and to defend your decision.

Chapter 9
Introductory Geometry

"Geometry and spatial sense are fundamental components of mathematics learning. They offer ways to interpret and reflect on our physical environment and can serve as tools for the study of other topics in mathematics and science. . . .(students) should develop clarity and precision in describing the properties of geometric objects and then classifying them by these properties into categories such as rectangle, triangle, pyramid, or prism. They can develop knowledge about how geometric shapes are related to one another and begin to articulate geometric arguments about the properties of these shapes."
—*Principles and Standards for School Mathematics*

"*Principles and Standards* calls for geometry to be learned using concrete models, drawing, and dynamic software. With appropriate activities and tools and with teacher support, students can make and explore conjectures about geometry and reason carefully about geometric ideas."
—*Principles and Standards for School Mathematics: An Overview*

In his book, *A Mathematician's Delight,* the renowned mathematician, W. W. Sawyer wrote: "The best way to learn geometry is to follow the road which the human race originally followed:

> Do things, arrange things,
> Make things, and only then
> Reason about things."

The activities in this chapter provide a variety of opportunities to arrange, measure, and construct geometric shapes in two and three dimensions. Following the constructions, you will conjecture and reason about the new figures to develop an understanding of their fundamental properties.

You will use pattern blocks and geoboards to explore angle measure and to discover the relationship between the number of sides of a convex polygon and the sum of the measures of its angles. Using a computer with the dynamic capability of geometric drawing software will assist you to develop a conceptual understanding of triangles and selected quadrilaterals, their properties, and the relationships among triangles and among the quadrilaterals.

Cubes will be used to help you develop spatial perception as you construct three-dimensional models of buildings based on various two-dimensional views. Constructing various two-dimensional nets and then attempting to fold them into polyhedra will strengthen your spatial visualization skills. The final activity in the chapter will introduce you to the famous Four-Color Map problem, one that is faced by cartographers everyday.

This informal exploration of shapes and their properties builds the foundation that is needed for the study of formal-deductive geometry.

Activity 1: What's the Angle?

PURPOSE	Develop the concept of measurement of angles using the central angle in a circle and its intercepted arc.
MATERIALS	A circular geoboard on pages A-39 and A-40 and geobands
GROUPING	Work individually.
GETTING STARTED	The number of degrees in an angle is the measure of the opening between the two rays or the amount of rotation as one ray turns from coinciding with one side of the angle to coinciding with the other side. A complete rotation of a ray about a point results in an angle with a measure of **360°**. One degree is $\frac{1}{360}$ of a complete rotation.

A **central angle** is an angle whose vertex is the center of a circle and whose sides contain radii.

Example:

For each central angle, record its measure and its classification: *acute, right,* or *obtuse.*

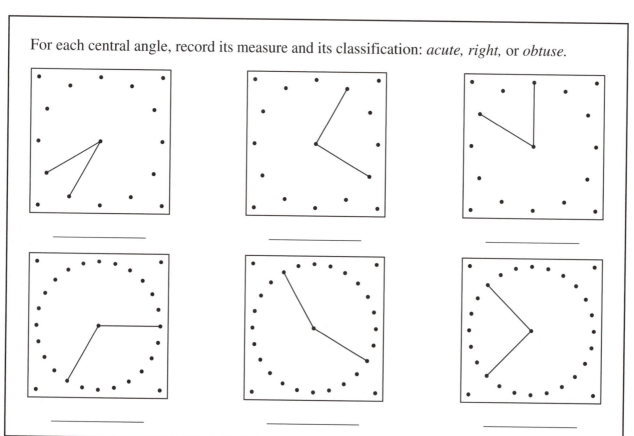

For each exercise, construct an angle that has the given measurement. In some exercises, one side of the angle is given.

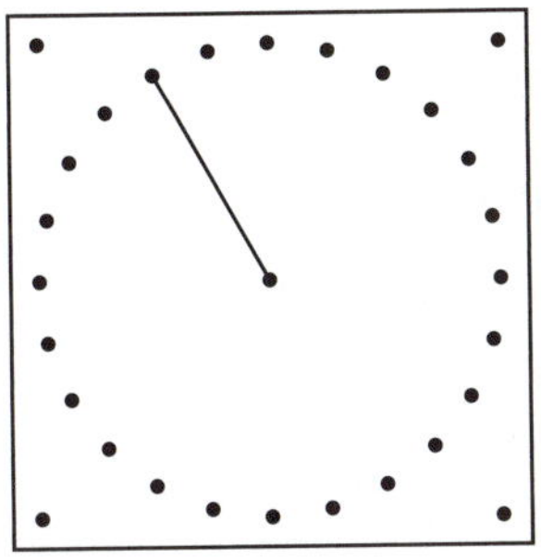

45°

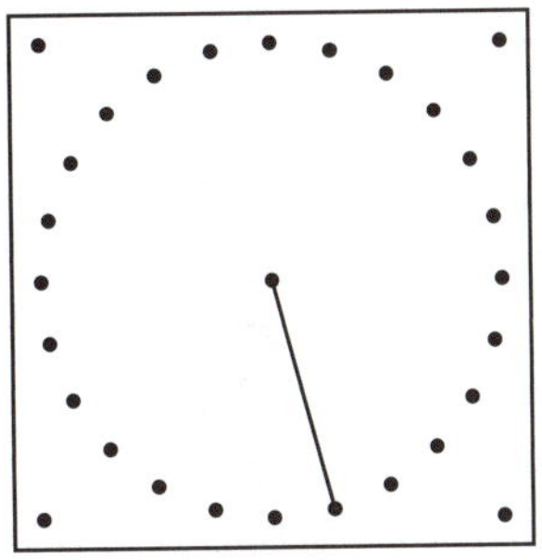

135°

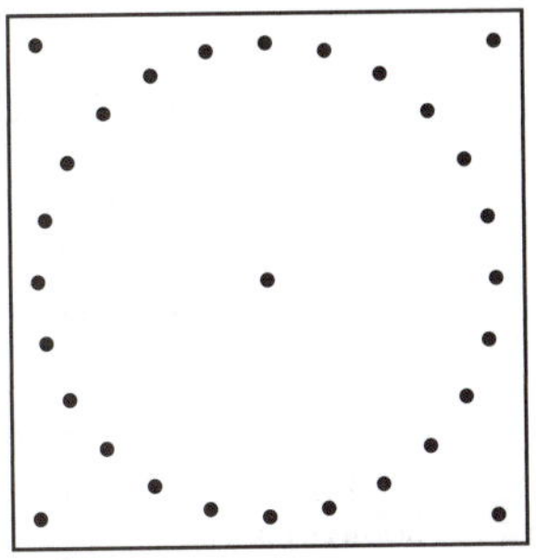

150°

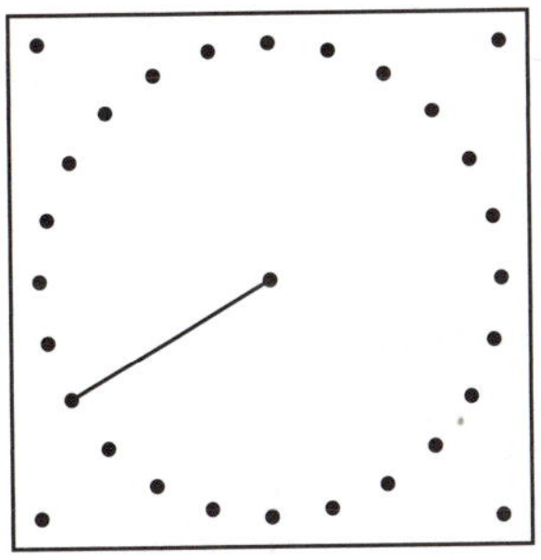

90°

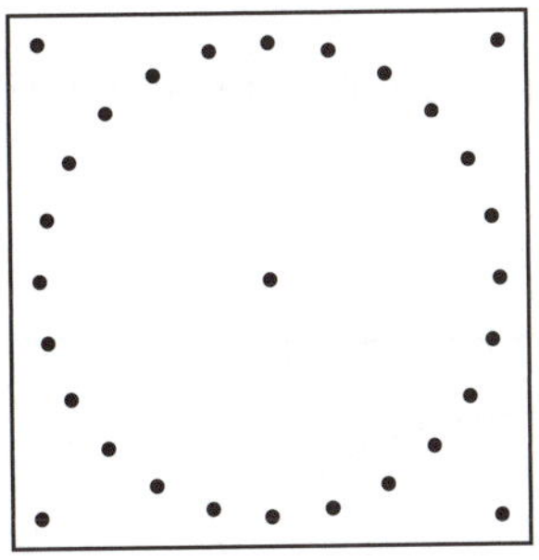

15°

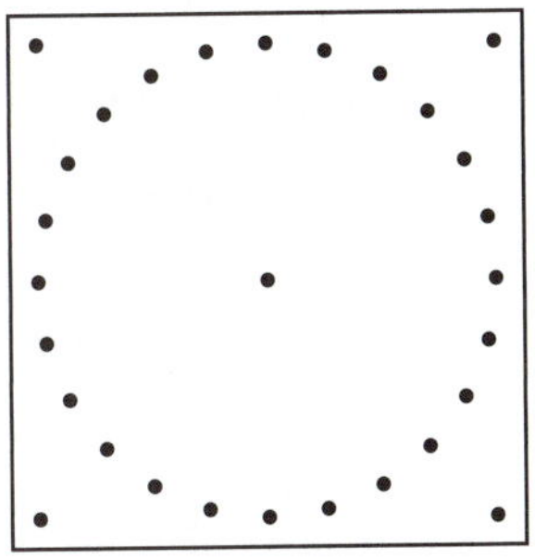

75°

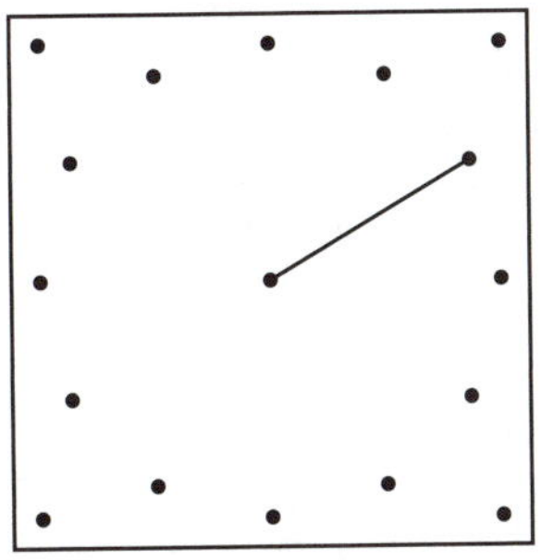

120°

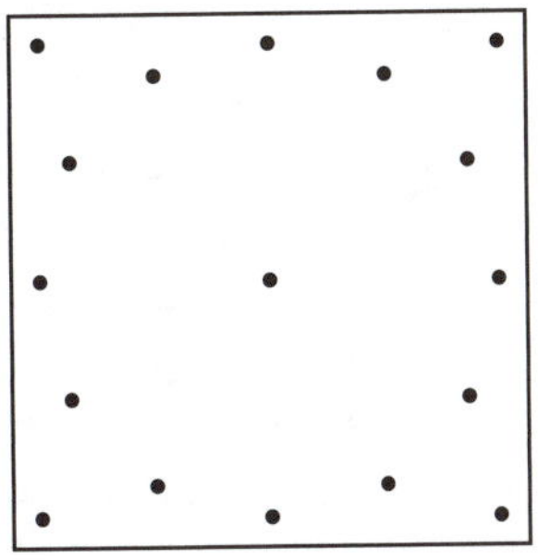

30°

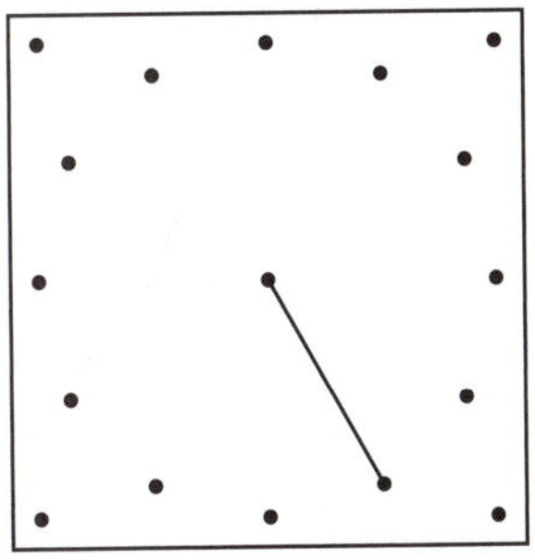

90°

Activity 2: Inside or Outside?

PURPOSE   Explore the relationship among points in the interior or exterior of a simple closed curve.

GROUPING   Work individually.

GETTING STARTED   The following examples illustrate various curves.

Examples:

Exterior   **Interior**

Simple
closed

Simple
not closed

Not simple
closed

Some *simple* curves may not appear to be so simple. In the curve shown below, point K is clearly outside the curve.

1.   Do you think point M is inside the curve or outside?

2.   What about point D?

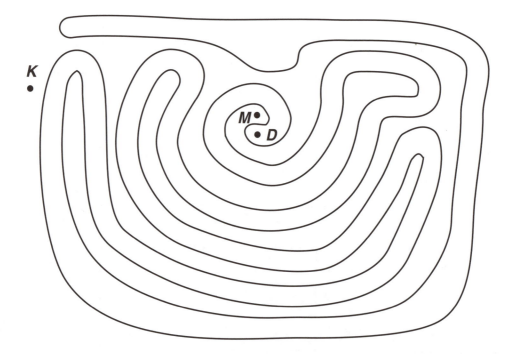

The answers to these questions may not be so obvious. Use the *simplify* problem-solving strategy to investigate the relationship between points inside and outside the simple closed curves on the following pages. Then return to answer the questions above.

For each of the following curves, point *M* and point *B* are either both inside or both outside the curve. Without crossing a boundary, draw a curve from *M* to *B*. Then draw a segment from *M* to *B* and count the number of times the segment crosses the boundary. (A *crossing* means that the segment goes from inside the curve to the outside, or from outside the curve to the inside.) Count the number of times the segment crosses the boundary and complete the table below.

Example:

Where are points M and B? _____

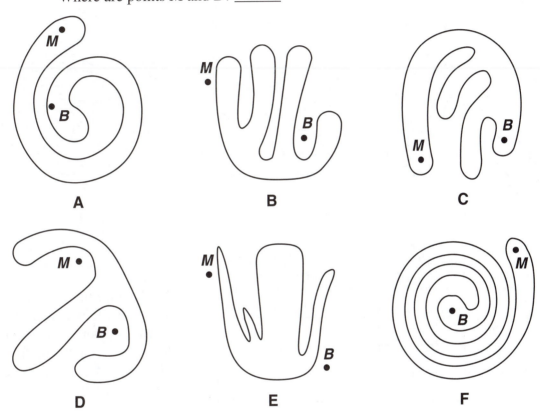

Figure	Number of Crossings	Number of Crossings Odd or Even
A		
B		
C		
D		
E		
F		

1. What conjecture can you make concerning the number of crossings when a segment is drawn between two points that are either both inside or both outside a simple closed curve?

In each of the curves, point *T* is inside the curve and point *B* is outside. Draw a segment from *T* to *B*, count the number of times the segment crosses the boundary, and complete the table.

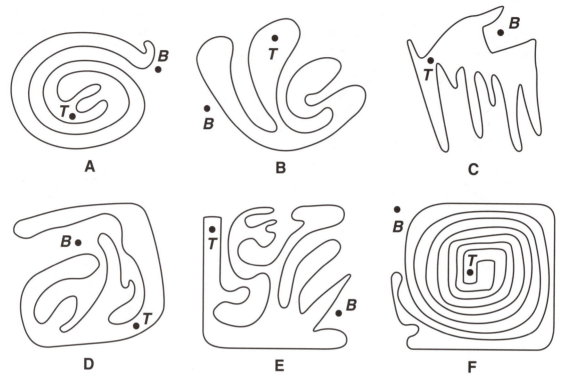

A   B   C

D   E   F

Figure	Number of Crossings	Number of Crossings Odd or Even
A		
B		
C		
D		
E		
F		

1. What conjecture can you make concerning the number of crossings when a segment is drawn between two points where one is inside and one is outside a simple closed curve?

2. Look back at the curve at the beginning of this activity. Explain how you can use the results of the activity to determine the correct location of points *M* and *D*.

3. If you were given a random point like point *M* at the beginning of the activity, explain how you could determine whether the point is inside the curve or outside. Test your conjecture by deciding if *M* is inside the curve or outside.

Activity 3: Triangle Properties—Angles

PURPOSE   Reinforce the theorem on the angle sum of a triangle and reinforce the construction of a triangle using a protractor and ruler.

MATERIALS   A centimeter ruler and a protractor

GROUPING   Work in pairs.

GETTING STARTED   Make all constructions on a separate piece of paper. As you construct the triangles in each section, compare your triangle with your partner's triangle.

Use a ruler and a protractor to construct a triangle that has two angles with the indicated measures. Record your results in the table.

1. 30°, 50°   2. 40°, 50°   3. 90°, 95°
4. 60°, 60°   5. 110°, 70°   6. 80°, 80°

Problem	Sum of the Given Angles	Is a Triangle Possible?	If Yes, What Is the Measure of the Third Angle?
1.			
2.			
3.			
4.			
5.			
6.			

1. In each Exercise 1–6 that **did** result in a triangle, what is true about the sum of the measures of the two given angles?

2. In each Exercise 1–6 that **did not** result in a triangle, what is true about the sum of the measures of the two given angles?

3. List the measures for three pairs of angles that can be used to construct a triangle.

   a. _____   b. _____   c. _____

4. List the measures for three pairs of angles that cannot be used to construct a triangle.

   a. _____   b. _____   c. _____

5. What can you conclude about the sum of the measures of the three angles of a triangle?

6. For each exercise that resulted in a triangle, list the angle pair and identify the type of triangle that was constructed.

Activity 4: Angles on Pattern Blocks

PURPOSE	Determine the sum of the measures of the interior angles of a polygon and the measure of each angle of a regular polygon.
MATERIALS	One set of Pattern Blocks (page A-7–A-11)
GROUPING	Work individually.

Determine the measure of each interior angle of each pattern block. You may use **only** the fact that the square has four right angles. **HINT:** You may place combinations of blocks on top of a block to assist in determining the measures of the angles.

Example:

Indicate the measure of each angle inside the pattern blocks as shown in the example. For each pattern block, explain the method you used to determine the measure of each angle. Draw a sketch of the blocks you used to illustrate your explanations. The measures you find on one block may be used to determine the measures of the angles on other blocks.

1.

2.

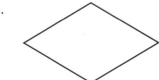

3.

4.

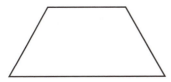

5.

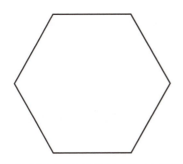

1. Use pattern blocks to construct a convex pentagon. Determine the measure of each interior angle of the pentagon. Sketch your pentagon and indicate the measure of each angle.

2. Use pattern blocks to construct a convex heptagon (seven sides). Determine the measure of each interior angle of the heptagon. Sketch your heptagon and indicate the measure of each angle.

3. Use the measures of each of the interior angles of the pattern blocks and the polygons you constructed to complete the following table.

Number of Sides	Sum of the Measures of the Angles
3	
4	
5	
6	
7	
8	
9	
n	

4. What is the relationship between the increase in the **number of sides** and the increase in the **sum of the measures** of the interior angles?

5. Given the number of sides of a polygon, n, how would you determine the **sum** of the measures of the interior angles of the polygon?

6. If a polygon is **regular**, how would you determine the measure of **each** interior angle?

Activity 5: Sum of Interior/Exterior Angles

PURPOSE Develop the relationships between the number of sides of a convex polygon and the sums of the measures of the interior and the exterior angles.

MATERIALS The Geometer's Sketchpad® or other geometry drawing software

GROUPING Work individually.

Use the geometry software to write a script to construct a convex polygon with five sides. Measure each interior angle and find the sum of the measures. Using the *drag* feature, pick one vertex and move it around to alter the figure.

Observe any changes in the measures of each angle and the **sum** of the measures of the interior angles. For one polygon, record the measure of each angle and the sum of the measures of the interior angles in the table.

Repeat the process for polygons with 6, 7, 8, and 9 sides.

Polygon	Measure of Each Interior Angle									Sum
	1	2	3	4	5	6	7	8	9	
5 sides										
6 sides										
7 sides										
8 sides										
9 sides										

1. What is the difference between successive sums in the table? _____

2. Write a rule to determine the sum of the measures of interior angles of a convex polygon given the number of sides (*n*).

Modify your script for constructing a polygon with five sides to construct the exterior angles as in the figure below. Begin at one vertex and continue clockwise; extend each side to form one *exterior* angle at each vertex.

Example:

Measure each exterior angle, and determine the sum of the measures. Using the *drag* feature, grab one vertex and move it around to alter the figure. Observe any changes in the measure of each exterior angle and the **sum** of the measures of the exterior angles. Record the measure of each angle and the sum of the measures of the exterior angles for one polygon in the table. Repeat the process for polygons with 6, 7, 8, and 9 sides.

Polygon	Measure of Each Interior Angle									Sum
	1	2	3	4	5	6	7	8	9	
5 sides										
6 sides										
7 sides										
8 sides										
9 sides										

1. What can you conclude about the sum of the measures of the exterior angles of a convex polygon? Explain.

EXTENSIONS

1. What is the sum of the measures of an interior angle of a convex polygon and one of its adjacent exterior angles? _____

2. Complete the following for a convex polygon with (*n*) sides.

   a. The sum of the measures of the interior angles and one adjacent exterior angle at each vertex, is _____.

   b. The sum of the measures of the interior angles is _____.

3. Use the results in Exercise 2 to prove algebraically that the sum of the measures of the exterior angles of a convex polygon, one at each vertex, is 360°.

Activity 6: Name That Polygon

PURPOSE Reinforce geometric vocabulary and the properties of polygons.

MATERIALS A length of rope or heavy string (2 to 3 m) tied at the ends to form a large loop

GROUPING Work in groups of four.

GETTING STARTED Three people in the group should each hold the rope at two different points. The fourth person reads the description of the polygon. The group then decides how each person must slide his or her hands along the rope to form the polygon. Once the group has correctly formed the polygon and decided on the correct name, each person should name the polygon and make a drawing of it at the right of the statement. Find as many different kinds of polygons as possible for each description. Switch roles after each problem.

---

1. A four-sided polygon with exactly two right angles.

2. An equilateral quadrilateral.

3. An equilateral quadrilateral with one right angle.

4. A polygon with at least two pairs of parallel sides.

5. A quadrilateral with congruent diagonals.

6. A quadrilateral with two pairs of congruent angles.

7. A polygon with one pair of parallel sides and two right angles.

8. A quadrilateral with two pair of congruent sides.

EXTENSIONS Each person in the group should write a statement that describes some polygon. Students in the group should then pick up the rope. The person who wrote the description should read it and the group should try to form the polygon and name it correctly.

Activity 7: Spatial Visualization

PURPOSE   Develop spatial visualization through exploration of polyhedrons and their corresponding polyhedral nets.

MATERIALS   Scissors, tape, graph paper, colored markers, Nets for the Open Top Box (page A-41), square tiles, a set of Polyhedra (optional), or a set of the Polyhedral Nets (pages A-42 and A-43)

GROUPING   Work in pairs or individually.

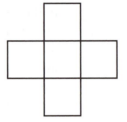

Make a copy of the nets for an open-top box.
Cut out a net, and fold it to form a box.

1. Unfold the net, cut off one square. Move it to a new position and tape it to the adjacent square to form a new net which will also fold into an open-top box. Record your answer on graph paper. Find all possible solutions.

2. Cut out another net for an open-top box and cut off any two squares. Move the squares to new positions; tape them to adjacent squares so that the new net will fold into an open-top box. (A net may **not** be congruent to a net formed in Exercise 1.) Record your answer on graph paper. Find all possible solutions.

When you complete Exercises 1 and 2, you should have seven new nets, no two of which are congruent, and all of which will fold into an open-top box.

Make copies of the original nets, cut them out, fold each into an open-top box and tape the edges to form eight boxes. Note that if you now cut the taped edges and unfold the box, the result is the net.

3. Pick one open-top box and one of the eight nets from Exercises 1 and 2. What edges should you cut along so that the box folds open to match the chosen net? Plan carefully. Mark the edges you plan to cut with a colored marker, then cut along the marked edges and unfold the box. If the unfolded box does not match the net you chose, refold the box, tape the edges, and try again.

4. Repeat the process for the remaining seven nets.

Draw the net shown at the right on
graph paper. Cut it out and fold it
to form a cube.

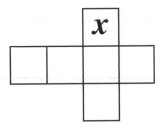

1. Cut off the square marked *x*, move it to a new position, and tape it to an adjacent square
   to form a new net that will also fold into a cube. Find all possible solutions so that no
   two nets are congruent.

2. When the net at the right is folded into a
   cube, what faces are opposite each other?
   Write three equations to show the sums of
   the pairs of numbers on opposite faces.

   _____ + _____ = _____   _____ + _____ = _____   _____ + _____ = _____

3. Write the missing digits on each net so that when you cut it out and fold it into a cube,
   the sums of the opposite faces of the cube are the same as those in Exercise 2.

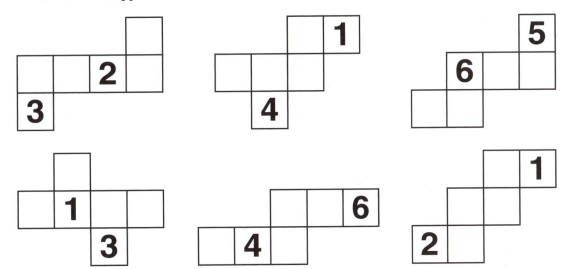

4. Desk calendars are made with number cubes where each face of a cube has one of
   the digits 0 through 9 on it. The cubes can be arranged so the faces will show the
   dates 01, 02, ... 31. What are the missing digits on each cube?

   Right cube _____, _____, _____

   Left cube   _____, _____, _____, _____

Each net in the left column will fold into a polyhedron. Use **Y** or **N** to indicate which of the nets to the right will also fold into the same polyhedron. If necessary, make the net with *Polydrons* or make copies of the nets on pages A-42 and A-43 and cut them out. Then check your conjecture by folding the net into the polyhedron.

1.

———   ———   ———

2.   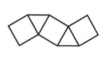

———   ———   ———

3.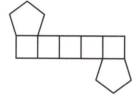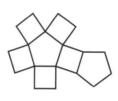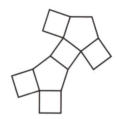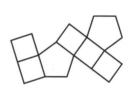

———   ———   ———

4.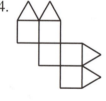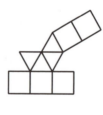

———   ———   ———

5.

———   ———   ———

6.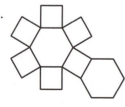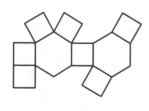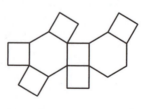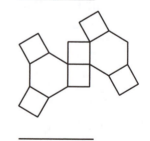

———   ———   ———

Activity 8: A View from the Top

PURPOSE   Develop spatial perception by using various views to construct models of buildings.

MATERIALS   Cubes, at least 16 per person or pair

GROUPING   Work individually or in pairs.

GETTING STARTED   Architectural plans may include various views of a building: top, front, back, left, and right. By viewing a building from the top and sides, you can determine its shape. Each number on a building mat tells you the number of stories (cubes) in that section of the building.

1. Use the numbers on this mat to construct the building with your cubes.

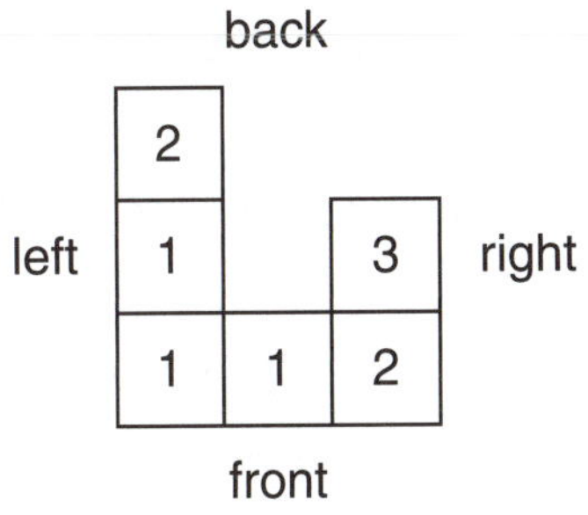

back

2	
left 1	3 right
1	1	2

front

2. Determine which views below represent the

front _____   back _____   right _____   left _____

A. 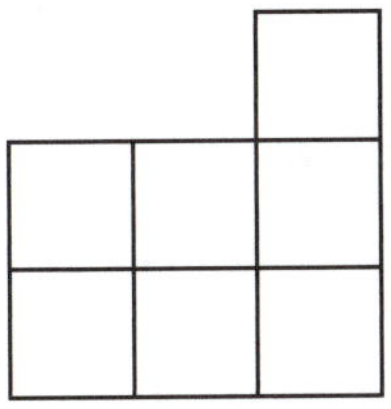   B. 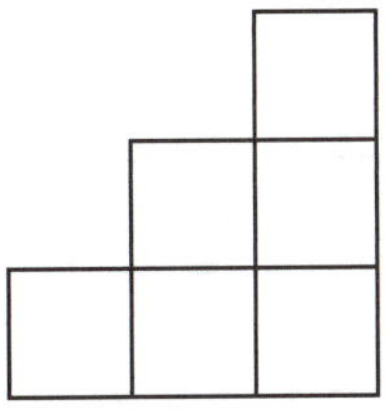   C.

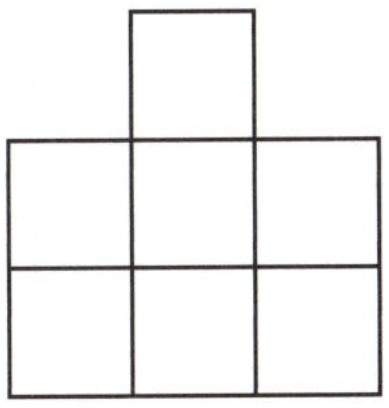

D. 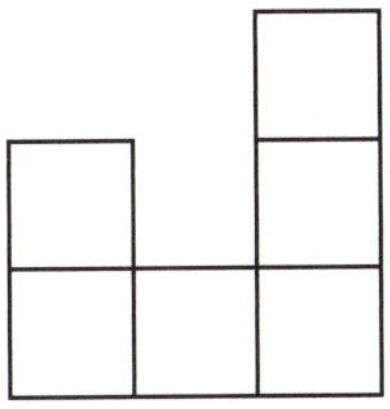   E. 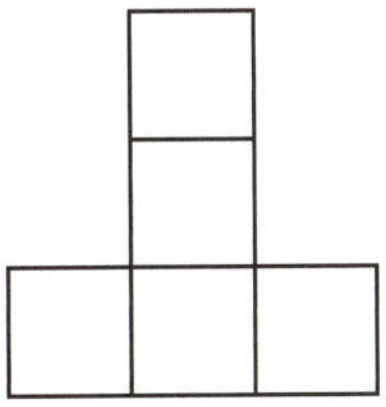   F.

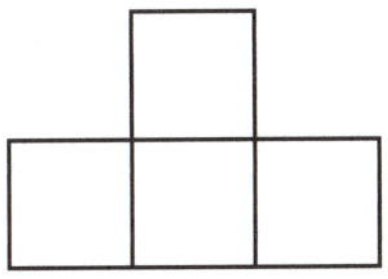

G. 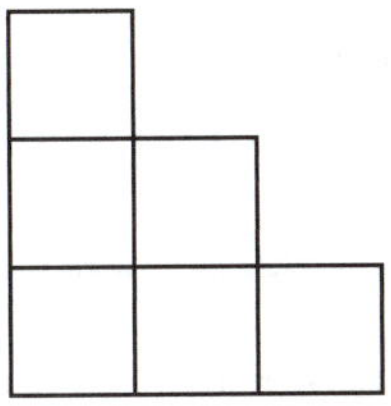   H. 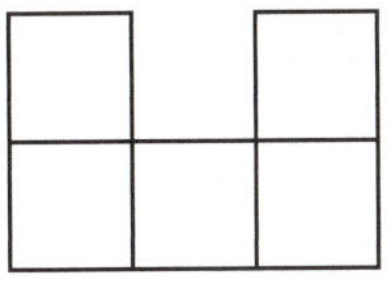   I.

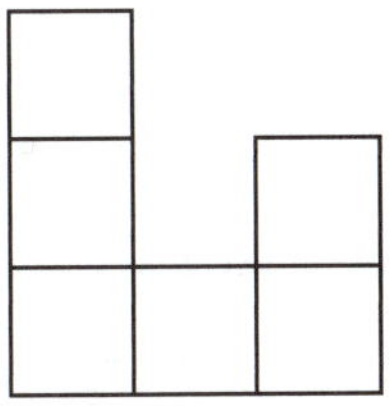

1. Use your cubes to construct the building represented by the following mats:

   a.
1	2	1
	3	
1	2	1

   Front

   b.
4	2	1
	2	3
	4	

   Front

   c.
3	2	
4	3	2
1		
1		

   Front

2. On centimeter grid paper, draw the architectural plans for each building. Label the top, front, back, left, and right view for each.

3. What is the relationship between the front and back views? Between the left and right views?

Use the plans below to construct each building.
Record the height of each section of the building on the mat.

BUILDING VIEWS

	TOP	FRONT	RIGHT	MAT

Example:

MAT
2	1	1
1		

1.

2.

3.

Activity 9: Map Coloring

PURPOSE   Determine the minimum number of colors needed to color all regions of a map so that no two regions that share a boundary are the same color.

MATERIALS   Colored pencils or crayons and blank United States map (page A-44)

GROUPING   Play the game with a partner and complete the activity individually.

GETTING STARTED   To play the Map Coloring Game, players alternate turns drawing a simple closed region and coloring it. Each new region must share part of a boundary with at least one of the regions previously drawn. No two regions that share any part of a boundary other than a point can have the same color. The first player to use the **fifth** color is the loser.

OK   NOT OK

---

1. Color the regions in Map A and Map B so that no two regions that share a boundary have the same color. Use the least number of colors possible.

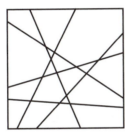

Map A

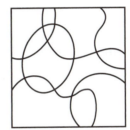

Map B

2. How many colors are needed? _____

---

1. Using the least number of colors possible, color the regions in Map C and Map D so that no two regions that share a boundary have the same color.

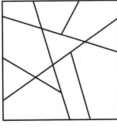

Map C

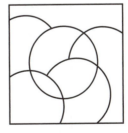

Map D

2. How many colors are needed? _____

3. How are Maps C and D different from Maps A and B?

1. Color the seven western states shown
    so that **no** two states that share a border
    have the same color. Use the least
    number of colors possible.

2. What is the minimum number of colors needed? _____

3. Why does this map require more colors than Maps A–D?

In each box below, construct a map that requires 2 colors, 3 colors, or 4 colors as indicated.
No two regions that share a boundary may have the same color.

2 colors	3 colors	4 colors

EXTENSIONS   1. You are required to color the map shown below according to the following rules. No two areas that share a border may be colored the same. Each region has an area of 8 square meters except the one marked 16. You have the following colors with enough paint to cover the area indicated. Indicate the correct color in each region.

Color	Coverage
Red	24 sq. m
Yellow	16 sq. m
Green	16 sq. m
Blue	16 sq. m

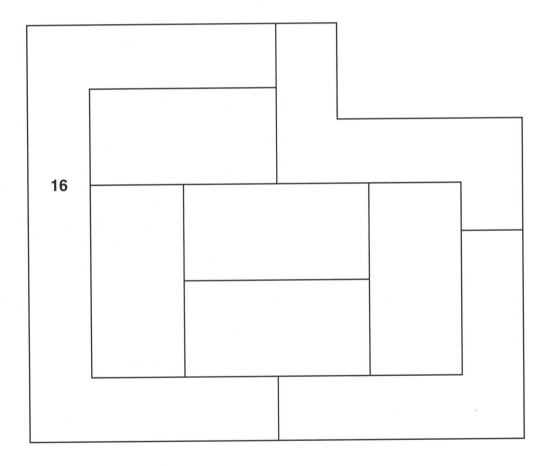

2. Find other groups of states on the blank United States map that require four colors. No two states that share a border may have the same color.

Chapter 9 Summary

Geometric shapes are a part of our everyday life. Throughout this chapter, the use of manipulatives allowed hands-on exploration of geometric figures and promoted the understanding of the basic concepts that are used in the study of geometry. The activities introduced you to several important geometric ideas.

In Activity 1, geoboards were used to explore angle measurement. Classifying the angles as acute, right, or obtuse laid the foundation for later exercises involving triangles. Linear and angle measurement was important in the development of the properties of both triangles and quadrilaterals.

Triangles and their properties were the subject of Activity 3. You were introduced to the angle sum of a triangle through construction using a protractor and ruler.

Pattern blocks and geometric drawing software were used in Activities 4 and 5 to develop the relationship between the number of sides of a convex polygon and the sum of the measures of the interior angles and the sum of the measures of the exterior angles. After using inductive reasoning to make a conjecture, you were asked to prove algebraically a theorem about the sum of the exterior angles.

In Activity 7, you investigated the arrangements of five squares in a net that could be folded into an open-top box—a transformation from two-dimensions to three-dimensions. You then had to cut along the edges of the box so that it would unfold into one of the nets—a transformation from three-dimensions to two-dimensions. Lastly, several nets for the same polyhedron were given and you had to guess which ones would fold into the polyhedron. You checked the accuracy of your guess by constructing each net and attempting to fold it into the given polyhedron.

Activity 8 helped develop spatial visualization in various ways. First, you had to determine two-dimensional views of a building as seen from different sides. Next, you were given the structure of a model and required to draw it as seen from four views. Finally, you had to construct a model of the building based on views from different sides. This activity has wide application in the real world in making architectural drawings and interpreting plans.

Activities 2 and 9 provided an opportunity to explore some different topics in geometry. Activity 2 included an exploration of points *inside* and *outside* a simple closed curve and Activity 9 introduced you to the famous Four Color Map Problem. A proof for this problem eluded mathematicians for centuries. It was finally solved in the late 1970's using computers.

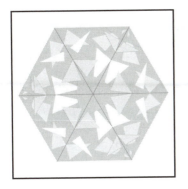

Chapter 10
Constructions, Congruence, and Similarity

". . .students also need experience in working with congruent and similar shapes. . . .they should understand that congruent shapes and angles are identical and can be "matched" by placing one on top of the other. Students can begin with an intuitive notion of similarity: similar shapes have congruent angles but not necessarily congruent sides. . . . they should extend their understanding of similarity to be more precise, noting for instance, that similar shapes "match exactly when magnified or shrunk" or that their corresponding angles are congruent and their corresponding sides are related by a scale factor."
—*Principles and Standards for School Mathematics*

The activities in this chapter will engage you in problem-solving situations in which you have the opportunity to explore and discover geometric concepts and relationships. You will construct a given triangle and compare the results with that of a classmate. Then, you will be asked to make and test conjectures based on your observations.

You will explore the concept of similarity using pattern blocks to construct and compare similar polygons. Then, you will use a protractor and ruler to construct triangles given certain sets of measures for the angles and check to determine if the triangles are similar. In two activities, you will use a computer drawing program to complete a construction and then alter the figure dynamically to illustrate that a conclusion is true for any triangle.

In the final activity, you will use the concepts that were developed in the previous activities on similar triangles to measure inaccessible heights. This activity demonstrates the use of similar triangles in a real world application. It provides a good example for the age old question, "when are we ever going to use this?"

Activity 1: Triangle Properties—Sides

PURPOSE   Develop the Triangle Inequality, reinforce construction of a triangle using a compass and ruler, and introduce congruence of triangles.

MATERIALS   A centimeter ruler, a compass, a geoboard, and geo-bands

GROUPING   Work in pairs.

GETTING STARTED   Make all constructions on a separate piece of paper. As you construct the triangles in each section, compare your results with those of your partner.

1. Use a ruler to construct two different triangles in which one side is 6 cm long and another side is 9 cm long.

   a. Compare your triangles with your partner's triangles. What do you notice?

   b. How many different triangles could you construct given the lengths of two sides?

2. Use a ruler and a compass to construct triangles with sides of the following lengths.

   a. 9 cm, 7 cm, 5 cm   b. 7 cm, 7 cm, 10 cm   c. 8 cm, 8 cm, 8 cm

   d. 6 cm, 5 cm, 12 cm   e. 7 cm, 12 cm, 11 cm   f. 10 cm, 6 cm, 4 cm

3. Which sets of measures **did not** result in a triangle? Why is it impossible to construct triangles with sides of these lengths?

4. Which sets of measures **did** result in a triangle? In these cases, what is true about the sum of the lengths of the two shorter sides?

5. List measures for three sets of three lengths of sides that **can** be used to construct a triangle. Do not use a set in which all the lengths are equal.

   a. _____   b. _____   c. _____

6. List measures for three sets of three lengths of sides that **cannot** be used to construct a triangle.

   a. _____   b. _____   c. _____

7. What can you conclude about the lengths of the sides of a triangle?

1. For each set of three measures in Exercise 2 that resulted in a triangle:

   a. Classify the triangle using a combination of classifications, one by sides and one by angles.

   b. Compare each of your triangles to the corresponding triangle of your partner. What do you notice?

2. With your partner, decide on two additional sets of measures for three segments that **will** form a triangle. Each person should construct a triangle using these numbers. Compare the triangles as before. What do you notice?

How many unique triangles can be constructed on a 3 × 3 portion of a geoboard? Build your triangles on a geoboard and record the results on the 3 × 3 grids below. Classify each triangle that you construct. It may be necessary to put more than one triangle on a grid.

Activity 2: To Be or Not to Be Congruent?

PURPOSE   Develop the triangle congruence axioms using constructions.

MATERIALS   A compass, a protractor, and a ruler

GROUPING   Work in pairs.

GETTING STARTED   In Activity 1, you learned that two triangles are congruent if the three sides of one triangle are congruent respectively to the three sides of the other. Are there other combinations of corresponding parts of two triangles that will ensure that the triangles are congruent?

1. Use a compass and ruler to construct a triangle with the parts given in each exercise.

    a.

    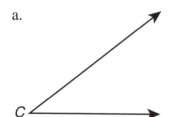

    b.

    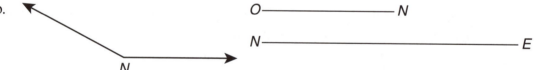

    c.

    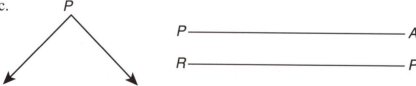

2. In each part of Exercise 1, how were the given sides and the angle of the triangle related?

3. Compare each triangle you constructed to the corresponding triangle of your partner. What do you notice?

4. What can you conclude from your answers to the above questions?

1. In the following problems, use a ruler and compass to construct a triangle with the given parts. Label the third angle of each triangle, *Z*.

a.

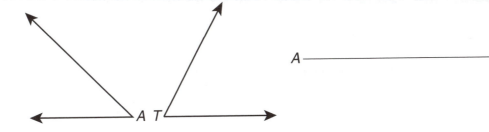

b.

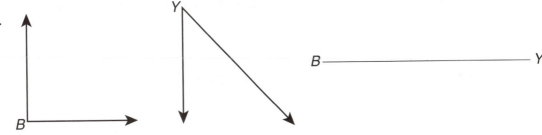

c.

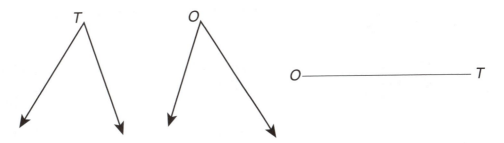

2. In each part of Exercise 1, how is the side related to the given angles?

3. Compare each of your triangles to the corresponding triangle of your partner. What do you notice?

4. What can you conclude from your answers to the above questions?

1. Use a ruler and a protractor to construct a triangle *BAD* in which *AB* = 10 cm, $m \angle A = 40°$, and $m \angle D = 80°$.

    a. How are the angles and the given side of the triangle related?

    b. Describe step by step how you constructed the triangle.

    c. Compare the triangle you constructed with that of your partner. What do you notice?

    d. What can you conclude from Parts (a) and (c)?

2. For each of the following, use a ruler, a protractor, and a compass to construct a triangle with the given parts.

    a.   *TO* = 8 cm
         OP = 5 cm
         $m \angle T = 30°$

    b.   *HA* = 10 cm
         AT = 6 cm
         $m \angle H = 37°$

    c.   *TI* = 9 cm
         IE = 4 cm
         $m \angle T = 40°$

3. For each part in Exercise 2, how are the given sides and angle related?

4. Which exercises did not result in a triangle?

5. Did any of the exercises result in more than one triangle? If so, which one(s)?

6. If any exercise resulted in only one triangle, what type of triangle was it?

7. What can you conclude from your answers in Exercises 3–6?

Activity 3: Paper Folding Construction

PURPOSE   Use paper folding to construct the perpendicular bisector of a line segment, the bisector of an angle, and the centroid, incenter, and circumcenter of a triangle.

MATERIALS   Squares of waxed paper (10–12 cm on a side) or paper squares used to separate hamburger patties, compass, ruler, protractor, and computer with geometry drawing software

GROUPING   Work in pairs.

GETTING STARTED   One student should read the directions and the other should do the folding. Both should review the results of the folding, answer any questions, and make a conjecture based on the findings. The conjecture should be compared with the conjecture of another pair of students and discussed so as to arrive at a mutually accepted conclusion. All of the drawing and folding should be done on the waxed paper.

1. Draw $\angle ABC$.

2. Fold $\overrightarrow{BA}$ onto $\overrightarrow{BC}$ and crease the paper.

3. Open the paper and mark a point D on the line formed by the crease.

4. a. Measure $\angle ABD$ and $\angle CBD$.
   b. What is true about the measures of $\angle ABD$ and $\angle CBD$?
   c. What is the relationship between $\overrightarrow{BD}$ and $\angle ABC$?

Recall that the distance from a point to a line is the length of the perpendicular segment drawn from the point to the line.

5. Place the edge of another piece of paper on side $\overrightarrow{BC}$ of $\angle ABC$ so that the adjacent side of the paper passes through point D as shown. Label the point E on $\overrightarrow{BC}$ and mark the length DE on the edge of the second sheet as shown. Then repeat the steps to find a point F on $\overrightarrow{BA}$. Compare DF and DE. What did you find? Place the point of your compass at D, open the compass to E, and draw a circle. What is the relationship between the circle and the sides of $\angle ABC$?

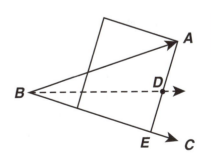

1. Draw a line segment $\overline{PQ}$.

2. Fold *P* onto *Q* and crease the paper.
   Open the paper and mark the point *M*
   as shown. *M* is the _____ of $\overline{PQ}$.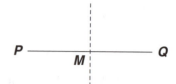

3. Mark a point *R* on the crease as shown.
   Measure $\angle RMP$ and $\angle RMQ$.
   What is the measure of each angle? _____

4. Describe the relationship between $\overleftrightarrow{RM}$ and $\overline{PQ}$.

5. Draw $\overline{RP}$ and $\overline{RQ}$. Fold the paper on *RM* again.
   What is true about *RP* and *RQ*? _____

6. Choose any other point *X* on $\overleftrightarrow{RM}$. What is true
   about *PX* and *QX*? _____

7. What can be said about any point *X* on $\overleftrightarrow{RM}$ and its relationship to *P* and *Q*?

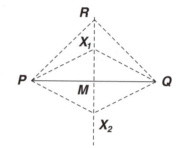

For the remaining constructions, you may use geometry software or paper folding. If you use
a drawing program, complete the construction and then alter the original triangle with the
dynamic feature of the program to verify that your conjecture works for *any* triangle.

1. Draw $\triangle ABC$ on your paper. Fold the paper to bisect
   each angle.

2. What appears to be true about the three angle bisectors?
   Check your results with those of another group.

3. Label the point where the bisectors intersect *P*.
   Place the edge of a second piece of paper along
   one side of the triangle as shown so that the
   adjacent side passes through *P*. Mark the point
   Q on $\overline{AB}$.

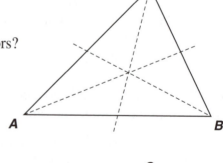

4. To determine if *P* is equidistant from the three sides
   of the triangle, place the point of your compass at *P*,
   open the compass to point *Q*, and draw a circle. The
   point *P* is called the *incenter* of the triangle.

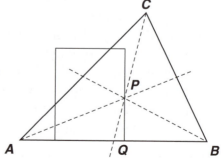

1. Draw an acute △*DEF*.

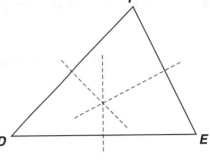

2. Fold the paper to construct the perpendicular bisector of each side of the triangle. What appears to be true about the perpendicular bisectors?

3. Mark the point of intersection *Q*. Measure *DQ*, *EQ*, and *FQ*. What did you find?

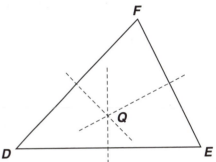

4. Place the point of your compass at *Q*, open it to *D*, and draw a circle. *Q* is called the *circumcenter* of the triangle.

5. Repeat the steps above for an obtuse triangle and a right triangle. How are the results the same? How are they different?

---

1. Draw △*XYZ*. Fold the paper to locate the midpoints of the sides of the triangle and label them *A*, *B*, and *C* as shown.

2. Draw the *medians* $\overline{XC}$, $\overline{YB}$, and $\overline{ZA}$. What appears to be true about the medians?

3. Mark the point of intersection, called the *centroid, M*. Measure the two segments on each median and find the ratio of the length of the shorter segment to the length of the longer segment. The *centroid* divides each median into two segments so that the ratio of their lengths is _____.

EXTENSIONS   Use geometry drawing software to construct the medians, angle bisectors, altitudes, and perpendicular bisectors of the sides of a triangle. The points at which these sets of line segments intersect are the *centroid, incenter, orthocenter,* and *circumcenter*, respectively. Research the Euler Line and find it in your triangle. Which of these points lie on the Euler Line? Using the dynamic feature of your software, verify that the Euler Line contains the same points in any triangle.

Activity 4: Pattern Block Similarity

PURPOSE Develop the concept of similarity using pattern blocks.

MATERIALS Set of Pattern Blocks (pages A-7–A-11)

GROUPING Work in pairs or individually.

GETTING STARTED Using squares, you can construct larger squares **similar** to the original one.

Example:

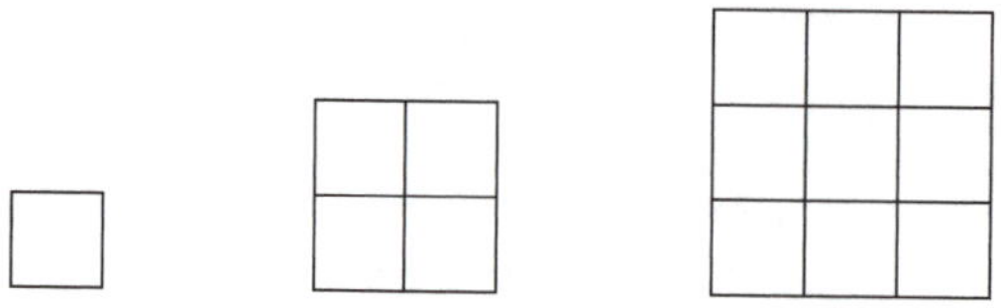

Figure 1   **Figure 2**   **Figure 3**

1. Use triangles to construct the next larger triangle similar to the green triangle. Sketch the similar triangle at the right.

2. For each pattern block, is it possible to use only blocks of the same shape to construct a larger shape similar to the original block? If so, construct the similar shape for each pattern block and sketch it below.

3. For each shape in Exercise 2, use the least number of blocks possible to construct the **next larger** similar shape. Sketch the figures below.

4. How many blocks were needed for each figure?

Repeat the process once more using more blocks. Record the number of blocks that are used to construct each similar figure for each pattern block in the table below.

Figure	1	2	3	4	5	6	7	8	n
Number of Blocks	1	4							

1. Describe the set of numbers in the **Number of Blocks** row of the table.

2. Describe the *numerical pattern of the differences* in the number of blocks in each successive similar figure.

Construct two different parallelograms using four red trapezoids each.

1. Are the measures of the angles of the red parallelograms equal to the measures of the corresponding angles of the blue rhombus?

2. Are the corresponding sides proportional?

3. Are the red parallelograms similar to the blue rhombus? Explain.

Construct a trapezoid using three red trapezoids.

1. Are the measures of the angles of the small trapezoid equal to the measures of the corresponding angles of the large trapezoid?

2. Are the corresponding sides proportional?

3. Are the trapezoids similar? Explain.

Construct a rhombus using four tan rhombuses. Compare this figure to the blue rhombus.

1. Are the corresponding sides of the two rhombuses proportional? If so, what is the ratio of the corresponding sides?

2. Are the measures of the corresponding angles equal?

3. Are the two rhombuses similar? Explain.

Activity 5: Similar Triangles

PURPOSE Introduce the concept of similarity of triangles and reinforce the construction of a triangle using a protractor and ruler.

MATERIALS A centimeter ruler and a protractor

GROUPING Work in pairs.

GETTING STARTED Make all constructions on a separate piece of paper.

1. Use a ruler and a protractor to construct a triangle that has three angles with the indicated measures.

   a. 35°, 55°, 90°   b. 40°, 75°, 65°   c. 130°, 25°, 25°

   d. 60°, 60°, 60°   e. 110°, 50°, 20°   f. 70°, 70°, 40°

2. Where is the longest (shortest) side of the triangle in relation to the largest (smallest) angle in the triangle?

3. Compare each of your triangles with the corresponding triangles of your partner. What do you notice? Explain your answer.

Measure the lengths of the sides of each of your triangles. Then find the ratio of the length of each side of your triangle to the length of the corresponding side of your partner's triangle and record the ratios in the table below.

Problem	Shortest Side to Shortest Side		Middle Side to Middle Side		Longest Side to Longest Side	
	Fraction	Decimal	Fraction	Decimal	Fraction	Decimal
a						
b						
c						
d						
e						
f						

4. For each pair of triangles, what can you conclude about the ratios of the corresponding sides?

Activity 6: Outdoor Geometry

PURPOSE	Apply the properties of similar triangles to indirect measurement.
MATERIALS	A mirror, a 5- to 10-meter measuring tape, a straw, a small washer, and a 40-cm length of string or thread
GROUPING	Work in groups of four.
GETTING STARTED	Identify several tall objects to be measured. One student should make a sketch of the method of solution (see example) and record the data. Another student can provide the shadow or do the sighting, as in the mirror method. The other two students can do the measuring. Divide the tasks among the members of the group and switch roles as each sucessive object is measured.

SHADOW METHOD

Measure *NA*, the height of a person; *AD*, the length of the person's shadow; and *JI*, the length of the shadow of the object for which the height is being determined.

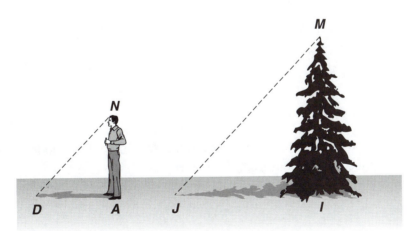

1. Explain why $\triangle DAN \sim \triangle JIM$.

2. Write a proportion that can be used to determine the height of the object, *MI*. Solve the proportion to find *MI*.

MIRROR METHOD

With a felt marker, draw a segment connecting the midpoints of one pair of opposite sides of a mirror. Place the mirror on the ground so that the segment on the mirror is parallel to a line determined by the tips of the toes of the shoes of a person facing the object to be measured. The person should look into the mirror and align the reflection of the top of the object to be measured (point J on the hoop) with the line on the mirror represented by $\overline{RS}$. Point M is the intersection of $\overline{AI}$ and $\overline{RS}$.

Measure *AM, MI,* and *EI.* Note that *EI* is the eye-to-ground distance, not the height of the person. The point I should be vertically below the person's eye (point E), approximately at the toes of the shoes.

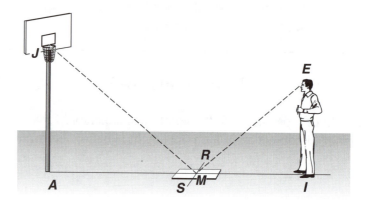

1. Explain why $\triangle JAM \sim \triangle EIM$.

2. Write a proportion that can be used to determine *JA.* Solve the proportion and find the height of the object being measured.

HYPSOMETER METHOD

Use a clipboard with a pad of paper attached to it. Pin a drinking straw along the top of the pad at C and B. Attach a small weight to one end of a 40-cm length of thread and tie the other end of the thread to the pin at B. One person should hold the clipboard and sight through the straw until the top of the object to be measured is sighted. A second person should mark the point F on the edge of the pad to determine the $\triangle BAF$. Measure DC, BA, AF, and the distance from eye level to the ground.

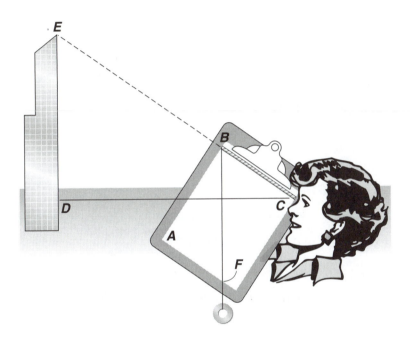

1. Explain why $\triangle CDE \sim \triangle BAF$.

2. Write a proportion that could be used to determine DE. Solve the proportion and find DE.

3. Is DE equal to the measure of the height of the object being measured? What must be done to DE to determine the height of the object?

Activity 7: Side Splitter Theorem

PURPOSE   Use a computer and geometric drawing software to explore similar triangles and the Side Splitter Theorem.

MATERIALS   Computer and dynamic geometric drawing software

GROUPING   Work individually.

Construct $\triangle ABC$. Locate a point M on $\overline{AB}$.
Construct $\overline{MN} \parallel \overline{AC}$ and such that N is on $\overline{BC}$.

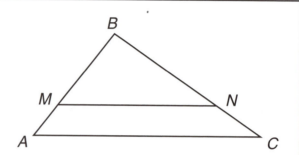

1. Measure BM, BN, BA, and BC.
    What is true about $\dfrac{BM}{BA}$ and $\dfrac{BN}{BC}$?

2. Drag $\overline{MN}$ to any position, M on $\overline{AB}$.
    Repeat Exercise 1.

3. What are the ratios when M is the midpoint of $\overline{AB}$?

4. a. Click on B and move point B to form a new $\triangle ABC$. What is true about
        $\dfrac{BM}{BA}$ and $\dfrac{BN}{BC}$?
    b. Drag $\overline{MN}$ to a new position. What is true about the ratios?

5. a. Measure MN and AC. How does $\dfrac{MN}{AC}$ compare to $\dfrac{BM}{BA}$ and $\dfrac{BN}{BC}$?
    b. Move B to form a new $\triangle ABC$ and repeat Part (a).

6. What conclusions can you make when a line segment intersects two sides of a triangle and is parallel to the third side?

7. What can you conclude about $\triangle ABC$ and $\triangle MBN$? Explain how you arrived at your answer.

Construct Δ *ABC*. Locate *X*, *Y*, and *Z*, the midpoints of $\overline{AB}$, $\overline{BC}$, and $\overline{AC}$, respectively. Locate *M, N,* and *P*, the midpoints of $\overline{XB}$, $\overline{BY}$, and $\overline{XY}$, respectively. Locate *D, E,* and *K*, the midpoints of $\overline{AX}$, $\overline{XZ}$, and $\overline{AZ}$, respectively. Locate *F, G,* and *L*, the midpoints of $\overline{ZY}$, $\overline{YC}$, and $\overline{CZ}$, respectively.

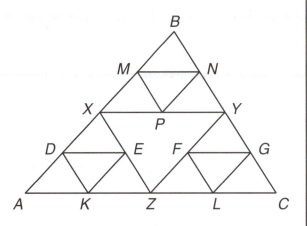

1. Use the Measure Menu to find the measurements for the segments and then determine the following ratios.

   a. $\dfrac{XY}{AC}$ b. $\dfrac{LG}{ZY}$ c. $\dfrac{DG}{AC}$

   d. $\dfrac{MN}{AC}$ e. $\dfrac{LG}{AB}$ f. $\dfrac{ZY}{NK}$

   g. $\dfrac{\text{perimeter of } \Delta\,XYZ}{\text{perimeter of } \Delta\,ABC}$ h. $\dfrac{\text{perimeter of } \Delta\,LGC}{\text{perimeter of } \Delta\,NKC}$

Click on point *B* and move it to form a new Δ *ABC*.

2. Compare the values of the ratios in Exercise 1 above to the ratios found in the new triangle. What did you find?

3. List three pairs of non-congruent similar triangles in the figure above. Explain how you determined that the triangles are similar. What is the ratio of the lengths of the corresponding sides in each pair?

4. List two pairs of congruent triangles. Explain how you determined that the triangles are congruent.

Activity 8: Mysterious Midpoints

PURPOSE   Apply coordinate geometry techniques to reinforce understanding of the properties of quadrilaterals.

MATERIALS   Graph paper, ruler, and a computer with a geometry software package

GROUPING   Work individually or in groups of 2 or 3.

1. Plot and label each of the following sets of points on a separate pair of coordinate axes. Draw four quadrilaterals by drawing segments connecting the points of each set in order.

   a. $P(2, 5), I(7, 2), N(12, 5), K(7, 8)$ b. $B(^-1, ^-5), R(4, 2), O(^-3, 7), W(^-8, 0)$

   c. $R(2, 2), O(^-3, ^-1), S(^-6, ^-5), E(^-4, ^-2)$ d. $P(9, 4), O(11, 8), L(2, 4), Y(0, 0)$

2. Mark the midpoints of the sides of each quadrilateral and label them consecutively M, A, T, and H. Connect the points in order to form another quadrilateral. What appears to be true about each of the polygons $MATH$?

3. Explain how you can use coordinate methods to check your conjecture.

1. Plot and label the following sets of points on separate coordinate axes. Connect the points in each set in order to form the following special quadrilaterals: a parallelogram, a rhombus, a rectangle, and a square.

   a. $D(4, 8), U(1, 3), C(10, 6), K(13, 11)$ b. $B(2, ^-2), I(9, 1), K(2, 4), E(^-5, 1)$

   c. $D(2, 3), A(^-3, 3), V(^-3, ^-4), E(2, ^-4)$ d. $P(^-2, ^-5), I(^-7, 0), C(^-2, 5), K(3, 0)$

2. Identify each of the quadrilaterals and explain how you determined your answer.

3. Locate the midpoints of the sides and label them consecutively M, O, N, and T. Connect the points in order to form quadrilaterals.

   a. Are any of the quadrilaterals $MONT$ that were formed special quadrilaterals? Explain how you determined your answer.

   b. Are any of the quadrilaterals $MONT$ the same type of quadrilateral as the original figure? Explain how you determined your answer.

4. In each of the quadrilaterals $MONT$ in Exercise 3, locate the midpoints of the sides, label them consecutively W, X, Y, and Z, and connect them in order to form new quadrilaterals. How are the new quadrilaterals related to the original figures drawn in Exercise 1?

1. Complete each of the following statements.

   a. If you connect the midpoints of the sides of a quadrilateral in order, the
   resulting figure is a _____.

   b. If you connect the midpoints of the sides of a parallelogram in order, the
   resulting figure is a _____.

   c. If you connect the midpoints of the sides of a rectangle in order, the
   resulting figure is a_____.

   d. If you connect the midpoints of the sides of a rhombus in order, the
   resulting figure is a _____.

   e. If you connect the midpoints of the sides of a square in order, the
   resulting figure is a _____.

2. Explain which properties of the original quadrilateral determine the special
   quadrilateral that is formed by connecting the midpoints of the sides.

EXTENSIONS

1. Use a geometric drawing program to construct a quadrilateral.
   Locate the midpoints of the four sides and connect them in order
   as you did to form the polygon *MATH*. Use the dynamic feature
   of the program to alter the quadrilateral in various ways to verify
   that your first conjecture was correct.

2. On separate screens, construct a trapezoid, parallelogram,
   rhombus, rectangle, kite, and square. In each figure, locate the
   midpoints of the sides and connect them in order. Then locate the
   midpoints of the sides of the new figure and connect them in
   order to form a second quadrilateral. Use the dynamic feature of
   the program to alter each of the original figures in various ways.

   Given that you start with a special quadrilateral, explain the
   relationship between the original quadrilateral and the
   quadrilateral that is formed by connecting the midpoints, and
   between the original quadrilateral and the second quadrilateral
   formed. Check your conjectures using the measuring capabilities
   of the software.

Activity 9: Rectangles and Curves

PURPOSE   Reinforce the concepts of perimeter and area. Display discrete and continuous data using graphs and develop the idea of a limit.

MATERIALS   Centimeter graph paper (page A-49), scissors, 36 colored squares (page A-7), and a loop of string 24 cm in circumference

GROUPING   Work individually or in pairs.

1. Determine all possible rectangles that can be constructed using the 36 squares. As you determine each rectangle, outline it on a piece of centimeter graph paper and label the base (*b*) and the height (*h*). Measure the length of the base, the height, and the perimeter (*P*) for each rectangle and record the measurements in Table 1. The length of the side of a square equals 1 unit.

TABLE 1: Area = 36

Base (*b*)									
Height (*h*)									
Perimeter (*P*)									

2. Cut out each rectangle. Construct a graph showing only the first quadrant. Label the horizontal axis (*b*) and the vertical axis (*h*). Place each rectangle on the axis as illustrated in the graph that follows. Each rectangle must be placed so that one vertex is at (0, 0). Make a drawing of the completed graph.

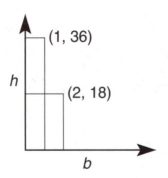

1. Using the data in Table 1, plot the ordered pairs (b, h) that represent the base and height of each rectangle on Graph 1.

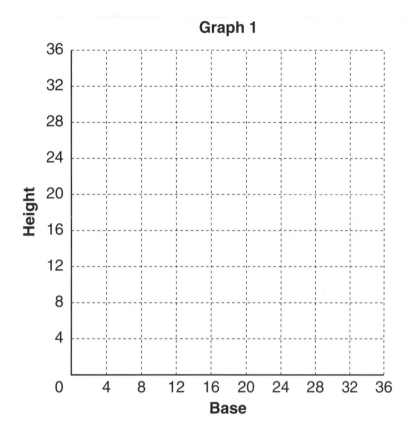

Graph 1

The data displayed in Graph 1 represents *all* rectangles that can be constructed using 36 squares.

2. Are there any other rectangles with an area of 36? Explain.

3. How many are there?

4. List the dimensions of at least three other rectangles that have an area of 36.

5. Are these rectangles represented in Graph 1? If not, describe where on the graph the new points should be placed.

1. On Graph 2, plot and connect the points that represent the ordered pairs (b, h) for *all* possible rectangles whose areas are 36. (**HINT:** Use the points on Graph 1.)

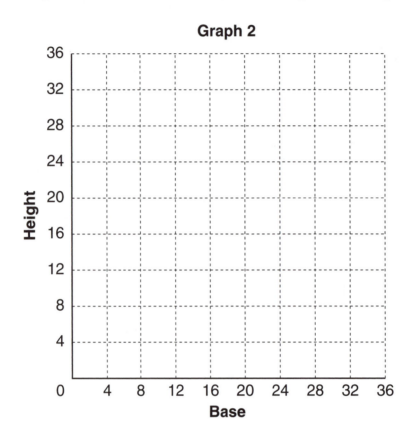

Graph 2

2. Will the graph of this data intersect either axis? If so, where? If not, explain.

3. Can any point on the graph lie below the horizontal axis? Explain.

1. Using the data from Table 1, plot the ordered pairs (b, P) on Graph 3. Connect the points on the graph with a smooth curve.

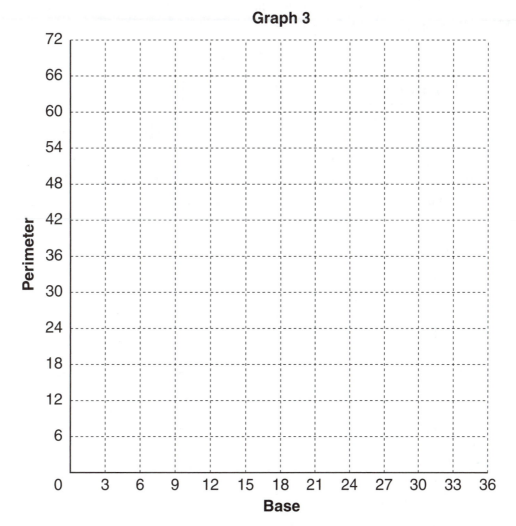

Graph 3

2. Each rectangle represented on the graph has an area of _____.

3. What is the least perimeter of any rectangle?

4. What are the dimensions of the rectangle with the least perimeter?

5. Is there a rectangle with a maximum perimeter? Explain.

6. The area of each rectangle represented on the graph is 36. Explain how it is possible for a rectangle to have an area of 36, yet have a perimeter that is unlimited.

7. Describe a physical model that you could use to illustrate the concept of a finite area being enclosed by an unlimited perimeter.

1. Determine all possible rectangles with integral dimensions and a perimeter of 24 cm
    that can be enclosed by the loop of string. Outline each rectangle on centimeter graph
    paper and label the base (*b*) and the height (*h*). Record the base (*b*), the height (*h*), and
    the area (*A*) for each rectangle in Table 2.

TABLE 2: Perimeter = 24

Base (*b*)										
Height (*h*)										
Area (*A*)										

2. Using the data in Table 2, plot the ordered pairs (*b*, *h*) that represent the base and height
    of each rectangle on Graph 4. Draw a line through the points.

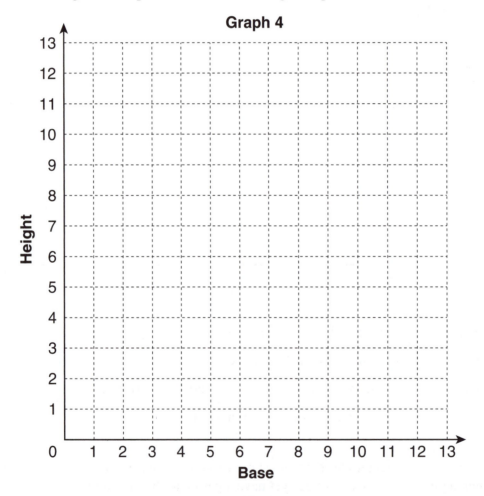

Graph 4

3. Graph 4 displays the data for *all* rectangles with a perimeter of 24 cm. Will the graph of
    this data intersect either axis? If so, where?

4. If the graph did intersect the horizontal axis, what would be the coordinates of that
    point? _____ How is the *b* coordinate of that point related to the perimeter?

1. Using the data from Table 2, plot the ordered pairs (*b*, *A*) on Graph 5. Connect the points on the graph with a smooth curve.

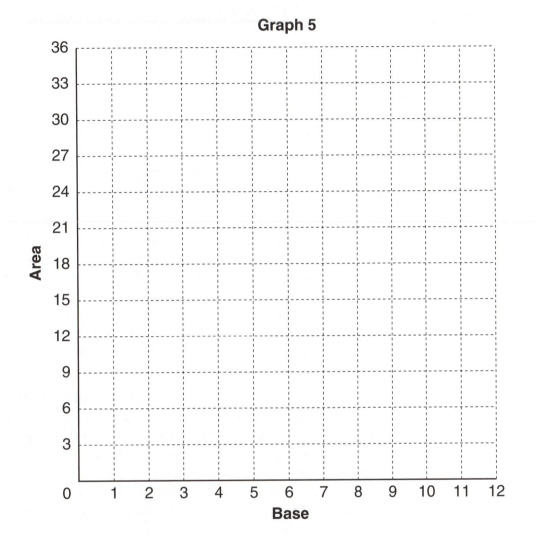

Graph 5

2. If a rectangle has an area of 24, what is the measure of the base?

3. What is the maximum area of any rectangle?

4. What are the dimensions of the rectangle with maximum area?

5. Is there a rectangle with a least area? Explain.

6. As the area of the rectangle approaches zero, the measure of the base approaches a maximum value of _____. Explain.

Chapter 10 Summary

In Activity 1, the Triangle Inequality was reinforced as you constructed triangles given the measures of their sides. When a triangle could be constructed, comparing your triangle with a classmate's introduced the idea that triangles are congruent if the corresponding sides are congruent.

Activity 2 developed the Side-Angle-Side (SAS), Angle-Side-Angle (ASA), and Angle-Angle-Side (AAS) axioms for congruence of triangles. When given the measures for two sides and an angle that is not included (SSA), you discovered that, depending on the given measures, three possibilities existed - one triangle, two triangles, or no triangle.

The Paper Folding activity developed concurrence theorems involving the angle bisectors, medians, altitudes, and perpendicular bisectors of the sides of a triangle. The incenter and circumcenter were introduced and you discovered that the circumcenter centroid and orthocenter lie on the Euler Line. By using dynamic geometry software, you were able to make the construction on one triangle and then alter the completed figure to show that the construction worked for many different triangles.

The concept of similarity was introduced with pattern blocks in Activity 4. Using the blocks, you could easily determine that the corresponding angles were congruent and that the ratios of the corresponding sides were equal. You also discovered that two polygons that have either corresponding angles congruent or the ratios of corresponding sides equal, may appear to be similar, but are not. Both conditions are necessary for the polygons to be similar.

When you constructed a triangle given the measures of the angles, you discovered through comparison of your triangle with a classmate's, that the triangles were not congruent, rather they were similar.

Geometry surrounds us in our world and applications of geometry abound. The Outdoor Geometry activity allowed you to apply the concepts developed in earlier activities in a real-world problem setting.

In Activity 8, you investigated the results when a new quadrilateral is formed by connecting the midpoints of the consecutive sides of a quadrilateral. You found that the type of quadrilateral formed was dependent on the original one. Connecting the midpoints of the consecutive sides of the new quadrilateral resulted in yet another quadrilateral, again related to the previous one and the original. The relationships between the quadrilaterals was easily determined using coordinate methods.

Coordinate geometry was used in Activity 9 to study the relationship between area and perimeter. You discovered that for a given area, there is a minimum perimeter that will enclose it, but no maximum perimeter. Thus it is possible to have a finite area bounded by an infinite perimeter. However, for a given perimeter, you found that it will enclose a maximum area, but no minimum area.

Chapter 11
Concepts of Measurement

"The study of measurement is crucial in the pre-K–12 mathematics curriculum because of its practicality and pervasiveness in so many aspect of everyday life. The study of measurement also provides an opportunity for learning about other areas of mathematics, such as number operations, geometric ideas, statistical concepts, and notions of function. It highlights connections within mathematics and between mathematics and areas outside of mathematics, such as social studies, science, art and physical education. . . . Whatever their grade level, students should have many informal experiences in understanding attributes before using tools to measure them or relying on formulas to compute measurements."

—Principles and Standards for School Mathematics

The activities in this chapter follow W.W. Sawyer's recommendation in *A Mathematician's Delight* to arrange things and make things before reasoning about them. Through exploration with a variety of manipulatives, you will develop an understanding of the concept of area and the formulas for finding the areas of certain polygons.

All of the formulas for area will be developed sequentially beginning with the relationship of the product of the dimensions of a rectangular array used to illustrate multiplication. Each new formula will be related to a previously developed one by comparing the actual areas of the two polygons. A similar sequence of activities will develop the rules for determining the volumes of certain polyhedra.

The Pythagorean Theorem, one of the most important theorems in all of geometry, will be explored informally through two puzzles. Each one involves manipulatives to clearly demonstrate the relationship between the area of the square constructed on the hypotenuse and the sum of the areas of the squares constructed on the other two sides.

In the final activity, you will explore the relationship between surface area and volume and apply this knowledge to the process of manufacturing boxes. As you collect and analyze the data, you will simulate an industrial application of geometry where decisions are based on minimizing costs and maximizing profits.

Activity 1: Units of Measure

PURPOSE   Develop an understanding of units of measure and the need for standard units.

MATERIALS   A handful of paper clips (all one size) and a pencil

GROUPING   Work in pairs to measure and then complete the activity individually.

Measure the items listed in the table and record your results to the nearest half unit. Answer all questions on a separate sheet of paper.

	Measure in number of paper clips	Measure in number of pencils
1. Width of your desk		
2. Length of teacher's desk		
3. Width of classroom door		

1. Were the measurements (number of paper clips, number of pencils) the same for all students? Explain.

2. If the width or length of each item is an *invariant,* why does the measure of an item differ?

3. During medieval times, the unit *foot* used for linear measure was based on the length of the king's foot. Over time, this caused many political, legal, and economic problems. Describe what these problems might have been.

4. Explain why it is necessary to establish a set of **Standard Units of Measure**.

EXTENSIONS   1. Research the history of units used to measure length, area, and volume. Name the unit (for example, foot) and explain the origin of the unit (for example, the length of the king's foot).

2. Recall the nursery rhyme *Jack and Jill.* Research the origin of this rhyme, its political implication at that time, and its relationship to units of measure.

Activity 2: Regular Polygons in a Row

PURPOSE   Reinforce the concept of perimeter and use the pattern problem-solving strategy to develop a rule for determining the perimeter of any number of regular polygons placed end to end.

MATERIALS   Pattern Blocks can be used for part of the activity.

GROUPING   Work individually.

1.  Perimeter = 3

2. 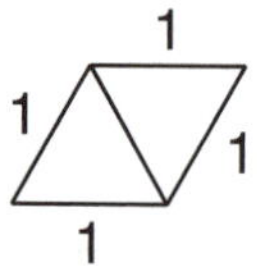 Perimeter = _____

3. 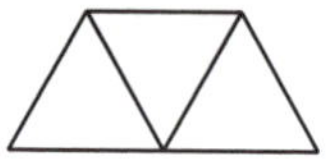 Perimeter = _____

4. 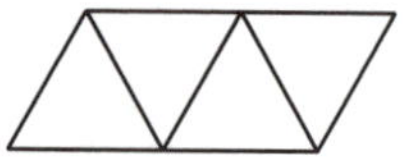 Perimeter = _____

5. What is the perimeter of 5 △s placed end to end as shown above? _____ 12 △s? _____

6. How many triangles would be needed to have a perimeter of 19? _____ 37? _____

7. Complete the following table.

Number of △s	1	2	3	4	5	9	13			28		100
Perimeter								19	23		35	

8. Given *k* triangles placed end to end, write a rule to determine the perimeter.

1. 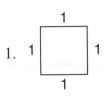   Perimeter = 4

2.    Perimeter = _____

3.    Perimeter = _____

4. What is the perimeter of 5 □s placed end to end? _____ 9 □s? _____

5. Complete the following table.

Number of □s	1	2	3							100
Perimeter										

6. What is the perimeter of 13 □s? _____ 23 □s? _____

7. How many squares would be needed to have a perimeter of 38? _____ 66? _____

8. Can the perimeter of k squares be 45? Why or why not?

9. Given k squares placed end to end, write a rule to determine the perimeter.

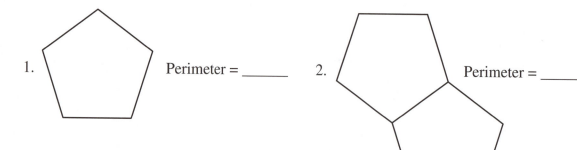

1.   Perimeter = _____

2.   Perimeter = _____

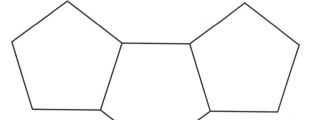

3.   Perimeter = _____

4.   What is the perimeter of 8 ⬠s? ____ 17 ⬠s? ____

5.   How many pentagons would be needed to have a perimeter of 65? ____

6.   What is the perimeter of 100 ⬠s? ____ Explain how you arrived at your answer.

7.   Can the perimeter of *k* pentagons be 49? Why or why not?

8.   Given *k* pentagons placed end to end, write a rule to determine the perimeter.

---

1.   Given *k* regular hexagons placed end to end, write a rule to determine the perimeter.

2.   Given *k* regular heptagons (seven sides) placed end to end, write a rule to determine the perimeter.

3.   Given *k* regular octagons (eight sides) placed end to end, write a rule to determine the perimeter.

Look for a pattern in the rules you developed to help write a general rule for finding the perimeter of k regular polygons with n sides that are placed end to end as in the previous exercises.

Regular Polygons	Number of Sides in Each Polygon	Rule
Triangles	3	$k + 2$
Squares		
Pentagons		
Hexagons		
Heptagons		
Octagons		
. . .		
n-gons	n	

Activity 3: What's My Area?

PURPOSE	Develop the concept of area through estimation and measurement using nonstandard units.
MATERIALS	One set of tangram pieces (page A-45)
GROUPING	Work individually or in pairs.
GETTING STARTED	Use the small triangles from the tangram set.

If the area of the triangle is one square unit, estimate the area of each figure below and on pages 214 and 215. Record your estimates. Then use the triangle to measure the area of each figure and record your answers.

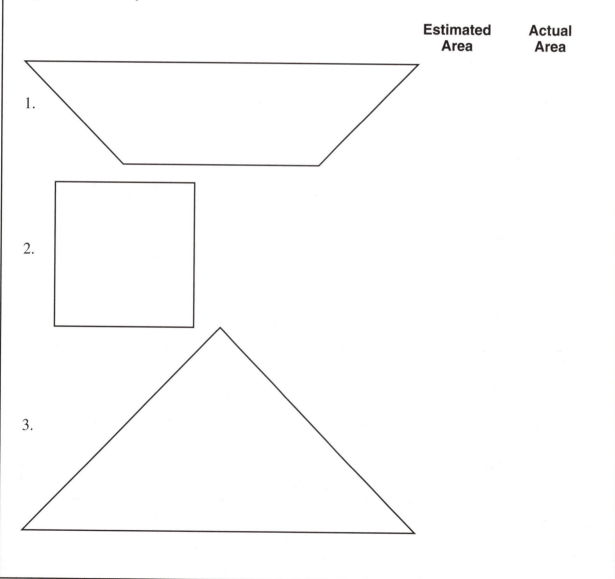

Estimated Area   **Actual Area**

1.

2.

3.

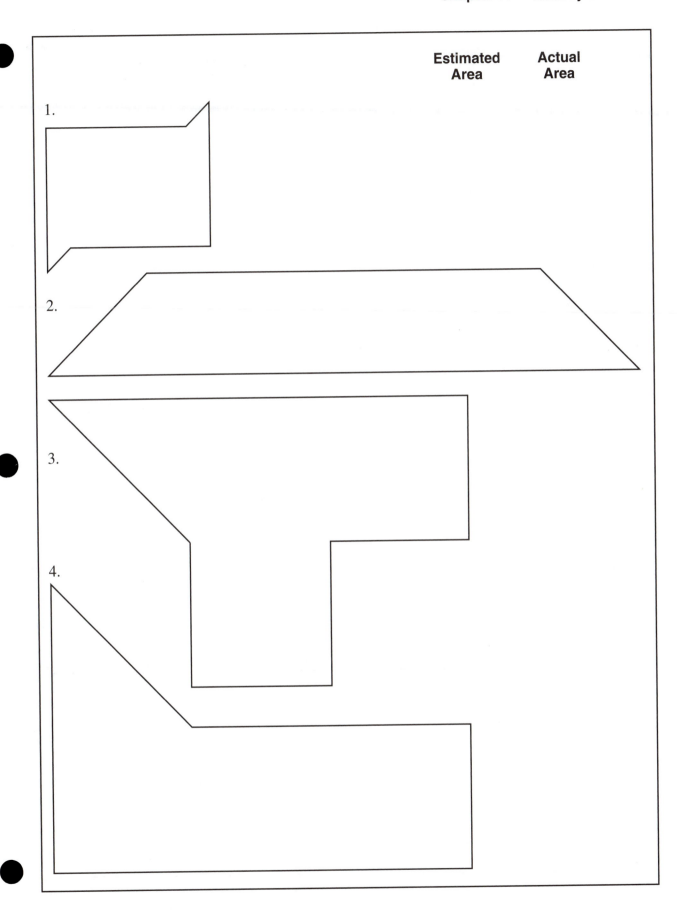

	Estimated Area	Actual Area
1.		
2.		
3.		
4.		

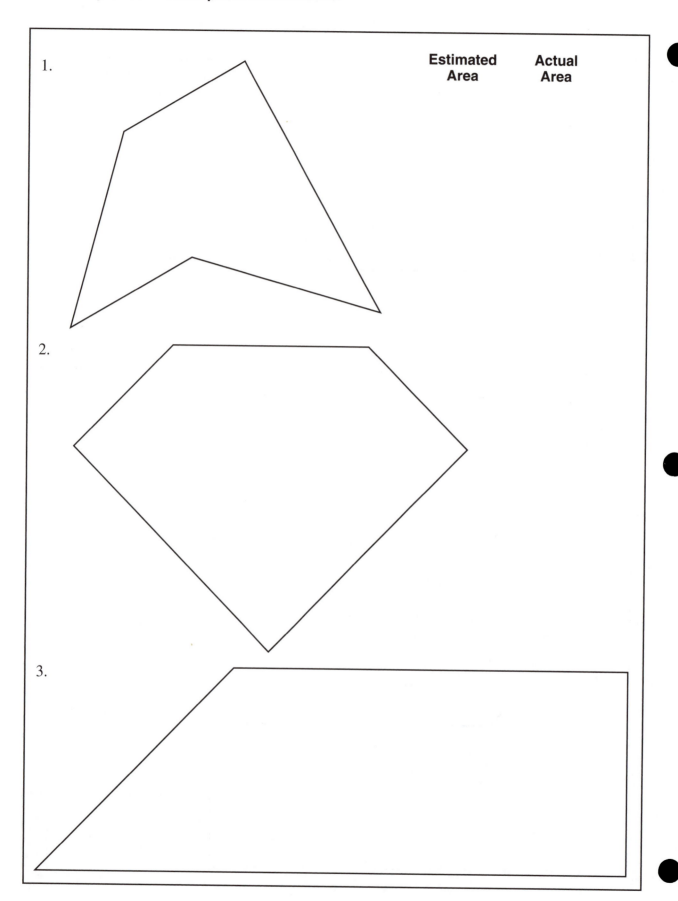

1.   Estimated   Actual
                                                        Area   Area

2.

3.

Activity 4: Areas of Polygons

PURPOSE	Develop the concept of the area of a polygon through the composition and decomposition of non-overlapping parts of a figure and reinforce the concept of convex and concave polygons.
MATERIALS	A geoboard, geobands, and dot paper (page A-50) for recording
GROUPING	Work individually.
GETTING STARTED	The smallest square that can be constructed on your geoboard has an area of one square unit.

---

1. On your geoboard, construct **10 or more** polygons that have an area of $2\frac{1}{2}$ square units and record your results on dot paper. Carefully check to be sure that each new figure is not congruent to a previous figure.

2. Sort the figures into sets of concave and convex polygons.

---

1. On your geoboard, construct **10 or more** polygons with an area of 3 square units and 10 or more polygons with an area of $3\frac{1}{2}$ square units. Record your results on dot paper. Carefully check to see that each new polygon is not congruent to a previous figure.

2. Sort the figures as above.

3. Look at the figures that you constructed with an area of 3 square units. Are the perimeters of all of the polygons equal? **NOTE:** The distance between two adjacent horizontal or vertical pegs is one unit. The distance between any other pair of pegs is more than one unit.

   a. What is the smallest perimeter?

   b. What is the greatest perimeter (approximately)?

   c. If you had a 1000 × 1000 peg geoboard, could you construct a polygon with an area of 3 square units and a perimeter of 100? 1000? Explain.

Activity 5: Pick's Theorem

PURPOSE Develop the concept of area, apply the patterns problem-solving strategy to determine a rule for finding the area of a polygon constructed on dot paper or a geoboard, reinforce the concept of dividing a figure into smaller non-overlapping parts to determine its area, and reinforce the concept of conservation of area.

MATERIALS A geoboard, geo-bands, and dot paper

GROUPING Work individually or in pairs.

GETTING STARTED The area of the smallest square that can be constructed on the geoboard or dot paper is **one** square unit.

For each polygon, count the number of points (geoboard or dot paper) on the perimeter, find the area, and enter the results in the table below.

Area = 1

a.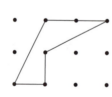

Area = _____

b.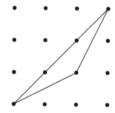

Area = _____

c.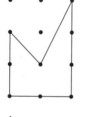

Area = _____

d.

Area = _____

e.

Area = _____

On your geoboard or on dot paper, construct several other polygons in which no points (geoboard or dot paper) are in the interior. Determine the area of each polygon and record the results in the table.

Number of Points on the Perimeter (N_p)	3	4	5	6	7			
Area of the Polygon (A)		1						

Look for a pattern in the table. Write a rule that relates the area of the polygon (A) to the number of points on the perimeter of the polygon (N_p).

PICK'S THEOREM REVISITED

On your geoboard or dot paper, construct three different polygons that have 1 point (geoboard or dot paper) in their interiors. Count the number of points (N_p) on the perimeter of each polygon and the number of points (N_i) inside the figure.

Determine the area (A) and record your data in the table below.

Construct several polygons that have 2 points in their interiors. Count the points as above and determine the area of each figure.

Continue this procedure, constructing polygons with 3 and 4 points in their interiors.

Determine the area of each polygon and record the data in the table.

Number of Points on the Perimeter (N_p)								
Number of Points in the Interior (N_i)								
Area of the Polygon (A)								

Examine the data in the table. Write a rule that relates the area of the polygon (A) to the number of points on the perimeter (N_p) and the number of points inside the figure (N_i).

Activity 6: From Rectangles to Parallelograms

PURPOSE   Develop a formula for finding the area of a parallelogram by comparing the area of a parallelogram to the area of a related rectangle.

MATERIALS   Dot paper (page A-50), a ruler, and scissors

GROUPING   Work individually.

Construct a parallelogram on your dot paper. Construct the altitude from one vertex of the upper base as shown. Cut out the parallelogram and then cut off the triangle. Move the triangle to the other end of the figure and match the vertices as shown.

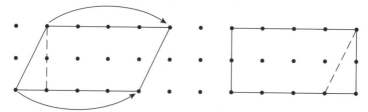

1. What kind of polygon is the new figure?

2. What is the relationship between the base and the altitude of the original parallelogram and those of the new polygon?

3. What is the area of the new polygon?

4. What is the relationship between the area of the original parallelogram and the area of the new polygon?

5. Describe two methods for finding the area of a rectangle.

Construct five additional parallelograms on your dot paper. Draw an altitude, cut out the figures, and construct a rectangle as shown above.

1. For each new parallelogram, determine the area of the related rectangle.

2. What is the relationship between the area of the parallelogram and the area of the related rectangle?

3. Write a rule to determine the area of a parallelogram.

Activity 7: From Parallelograms to Triangles

PURPOSE   Develop a formula for finding the area of a triangle by comparing the area of a triangle to the area of a related parallelogram.

MATERIALS   Dot paper (page A-50), a ruler, and scissors

GROUPING   Work individually.

Construct $\triangle KIM$ on your dot paper. Construct a segment $\overline{KE}$ that is parallel to $\overline{MI}$ and has length equal to the measure of $\overline{MI}$ as shown. Draw the segment $\overline{EM}$.

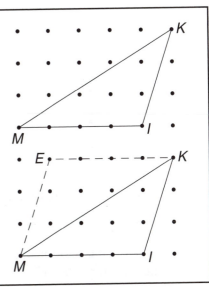

1. Polygon *MIKE* is what type of quadrilateral?

2. What is the area of quadrilateral *MIKE*?

3. The area of $\triangle KIM$ is what fractional part of the area of the quadrilateral *MIKE*?

4. What is the area of $\triangle KIM$?

Construct five additional triangles on your dot paper. Then construct the related parallelogram as shown above.

1. For each new triangle, what is the relationship between the area of the triangle and the area of the related parallelogram?

2. For each new triangle, what is the relationship between the base and altitude of the triangle and those of the related parallelogram?

3. What is the formula for finding the area of a parallelogram?

4. Use the relationship between the area of a triangle and the area of a related parallelogram to write a rule for determining the area of a triangle.

Activity 8: From Parallelograms to Trapezoids

PURPOSE   Develop a formula for finding the area of a trapezoid by comparing the area of a trapezoid to the area of a related parallelogram.

MATERIALS   Dot paper (page A-50), a ruler, and scissors

GROUPING   Work individually.

Construct a trapezoid on your dot paper as shown. Duplicate the trapezoid, rotate the copy, and match the vertices with the original trapezoid as shown.

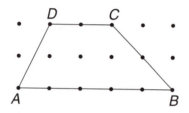 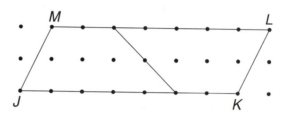

1. Polygon *JKLM* is what type of quadrilateral?

2. What is the area of quadrilateral *JKLM*?

3. What is the relationship between the area of trapezoid *ABCD* and the area of quadrilateral *JKLM*?

4. What is the area of trapezoid *ABCD*?

Construct five additional trapezoids on your dot paper. Then construct the related parallelogram.

1. For each trapezoid, what is the relationship between the area of the trapezoid and the area of the related parallelogram?

2. What is the relationship between the measure of the base of the parallelogram and the measures of the two bases of the trapezoid?

3. Use the relationship between the area of a trapezoid and the area of a related parallelogram to write a rule for determining the area of a trapezoid.

Activity 9: Pythagorean Puzzles

PURPOSE   Use puzzles to explore the Pythagorean Theorem.

MATERIALS   Scissors and a copy of the Pythagorean Puzzle sheet (page A-47)

GROUPING   Work individually.

PUZZLE I

1. Measure the three angles of triangle *XYZ*.

2. $\triangle XYZ$ is a _____ triangle.

   Cut out Square A and the four pieces of Square B. Put them together to form Square C.

3. What conclusion can you make about the sum of the areas of Square A and Square B as compared to the area of Square C?

PUZZLE II

- Cut out the five puzzle pieces in Puzzle II.
- Put the two triangles and the two pentagons together to form a square.
- Place this square on a sheet of paper. Trace around the perimeter; cut out the square and label it Square B.
- Now, put the original five pieces together to form a **larger** square.
- Place this square on a sheet of paper. Trace around the perimeter; cut out the square and label it Square C.

1. What conclusion can you make about the sum of the areas of Square A and Square B as compared to the area of Square C?

- Place Square A, Square B, and Square C together as in Puzzle I to form a triangle.

2. The triangle formed is a _____ triangle.

Activity 10: Right or Not?

PURPOSE Develop or reinforce the Pythagorean theorem and its converse and explore the relationship between the sides of a triangle and its classification as acute, right, or obtuse.

MATERIALS Centimeter graph paper (page A-49) and scissors

GROUPING Work individually or in pairs.

From a sheet of graph paper, cut out squares with areas 9, 16, 25, 36, 49, 64, 81, 100, 121, 144, and 169. Use three squares to construct a triangle as shown.

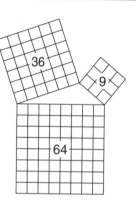

Determine if the triangle is acute, right, or obtuse. If necessary, compare the angles of the triangles to the corner of an index card to determine if the angles are acute, right, or obtuse. Enter the data in Table 1. Use additional sets of three squares to construct other triangles and enter the data in Table 1.

TABLE 1

Area of the Largest Square	Area of the Smallest Square	Area of the Third Square	Sum of the Areas of the Two Smaller Squares	Is the Triangle Acute, Right, or Obtuse?
64	9	36	45	obtuse
100	36	64		
36	16	25		
169				
121				
144				

Use the data from Table 1 to complete Table 2.

TABLE 2

Length of the Longest Side of the Triangle	Square of the Length of the Longest Side	Length of the Shortest Side of the Triangle	Length of the Third Side of the Triangle	Sum of the Squares of the Two Shorter Sides	Is the Triangle Acute, Right, or Obtuse?
8	64	3	6	45	obtuse

Use the data from Table 2 to complete the following statements.

a. If the square of the length of the longest side of a triangle is **less than** the sum of the squares of the lengths of the two shorter sides, the triangle is

    a(n) _____ triangle.

b. If the square of the length of the longest side of a triangle is **equal to** the sum of the squares of the lengths of the two shorter sides, the triangle is

    a(n) _____ triangle.

c. If the square of the length of the longest side of a triangle is **greater than** the sum of the squares of the lengths of the two shorter sides, the triangle is

    a(n) _____ triangle.

Activity 11: Now You See It, Now You Don't

PURPOSE	Explore the concept of conservation of area in a problem-solving setting.
MATERIALS	Graph paper, a ruler, a calculator, and scissors
GROUPING	Work individually.

1. On a piece of graph paper, draw an 8×8 square and divide it into four parts as shown in the figure at the right. What is the area of the square?

    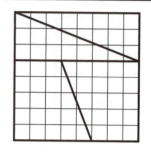

2. Cut out the four parts and form a rectangle as shown. What is the area of the rectangle? Can you explain this?

    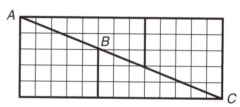

3. Rearrange the four parts as shown. What is the area of this figure? Is this possible?

    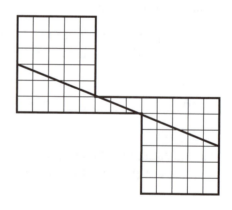

4. Use the Pythagorean Theorem to compare $AB + BC$ to AC. What did you find? What does it mean?

The Fibonacci numbers are found in a famous sequence of numbers: 1, 1, 2, 3, 5, 8, ... Each number in the sequence is the sum of the previous two numbers.

What are the next four Fibonacci numbers?

Note that the length of a side of the square was a Fibonacci number and that three sides were divided into two parts. The measure of each part was also a Fibonacci number.

1. On graph paper, draw a 13 × 13 square. Divide the square into four parts as shown. In this case, divide each side into two parts with measures 5 and 8. Note that, as in the previous exercise, the length of the side of the square (13) and the lengths of the two parts (5 and 8) are Fibonacci numbers.

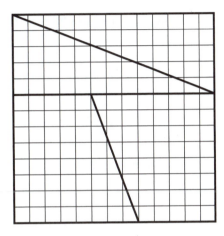

2. What is the area of the square?

3. Rearrange the four parts into a rectangle as in Exercise 2 on page 258. What is the area of the rectangle?

4. Rearrange the four parts into the figure shown in Exercise 3 on page 258. What is the area of this shape?

---

1. What is the next Fibonacci number after 13?

2. If you were to construct a square with that number as the length of the side, how would you divide the sides in order to make the four parts as in the previous exercises?

3. If you rearrange the four parts to form a rectangle and the other shape as before, what would be the area of the rectangle? _____ of the other shape? _____

EXTENSIONS   For any number in the Fibonacci sequence ($n \geq 3$), what is the relationship between the area of the square with that number as the length of a side and the areas of the rectangle and the other shape that can be constructed if the original square is divided and rearranged twice as was done in this activity?

Activity 12: Surface Area

PURPOSE	Develop the concept of surface area of solids.
MATERIALS	Scissors, tape, and the nets for solids (pages 261 and 262)
GROUPING	Work in pairs.
GETTING STARTED	Make a copy of the nets for the solids on construction paper. Cut out the nets. Then fold them and tape them together to make the solids.

Describe the geometric solid that is formed with each net.

1. Net A

2. Net B

3. Net C

4. Net D

Find the surface area of the solid formed with each net. Explain your method.

1. Net A

2. Net B

3. Net C

4. Net D

EXTENSIONS	Make three copies of Net B. Cut them out. Then fold them and construct the pyramids. Put the three pyramids together to form a prism. What is the volume of the prism? What is the volume of each pyramid?

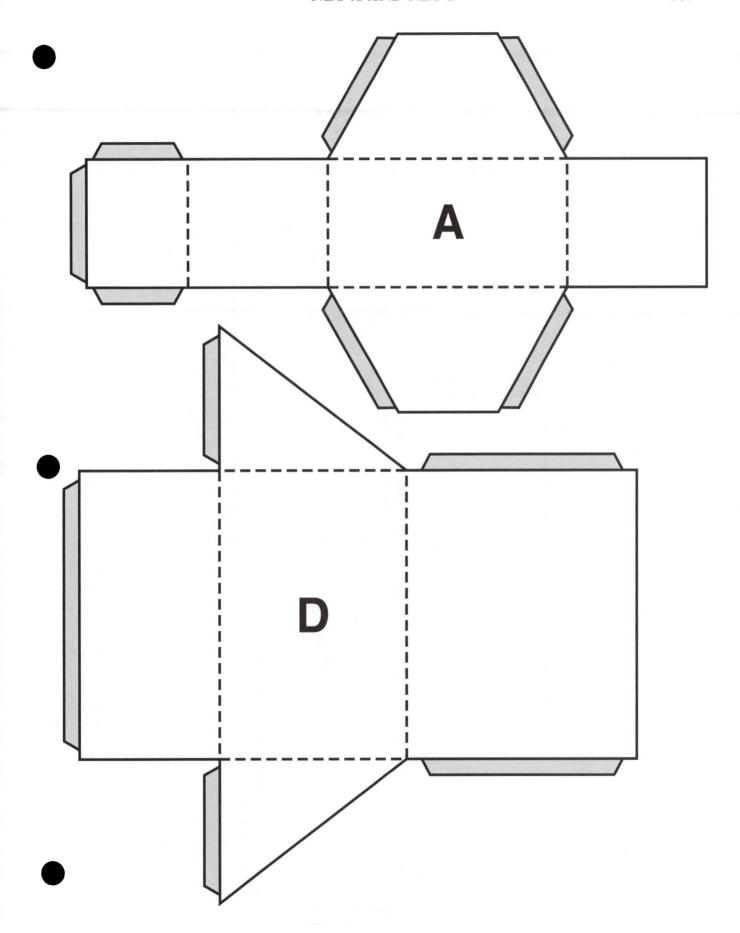

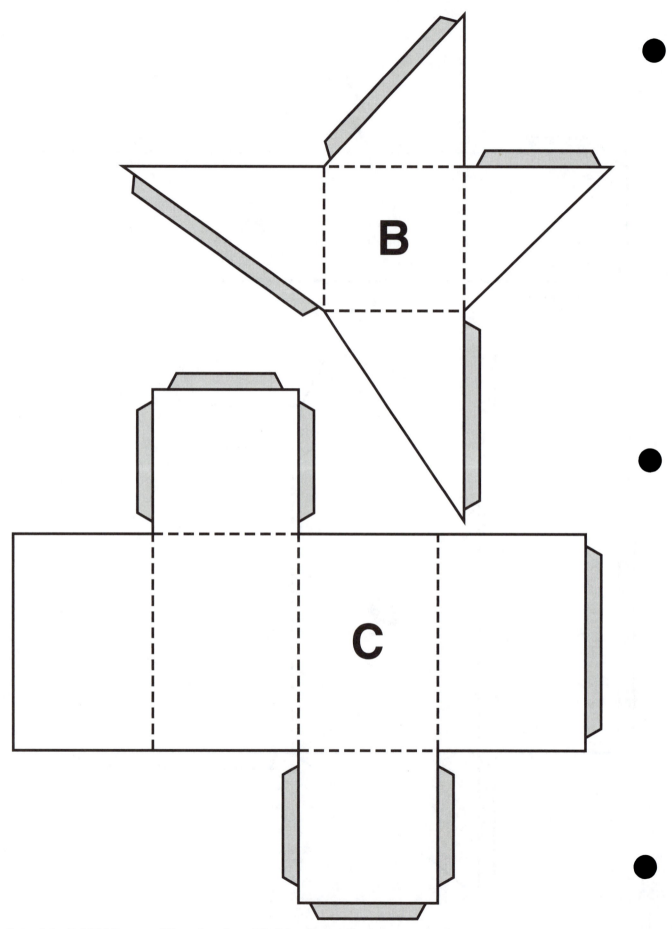

Activity 13: Volume of a Rectangular Solid

PURPOSE   Develop the formula for finding the volume of a rectangular prism.

MATERIALS   A set of cubes of the same size

GROUPING   Work individually.

GETTING STARTED   The numbers in the squares in each figure below indicate the number of cubes in a stack. Use your cubes to construct solids made up of these stacks of cubes.

4	3	1
2	1	5
1	2	2

Figure a

8	2
10	4
3	6

Figure b

4	3	6	5
2	5	2	3
4	3	6	4

Figure c

1. How many cubes are there in **each layer** of each solid and what is the total number of cubes in each solid?

   TOTAL

   a. __ __ __ __ __   _____

   b. __ __ __ __ __ __ __ __ __ __ __   _____

   c. __ __ __ __ __ __   _____

2. How many more cubes must be added to each solid to construct a rectangular prism? The base and height of the prism must be the same as the base and the maximum height of the solid.

   a. _____   b. _____   c. _____

3. What is the volume of each rectangular prism in Exercise 2 (the total number of cubes)?

   a. _____   b. _____   c. _____

Use the information in the table to construct rectangular prisms with the given dimensions and complete the table.

Length	Width	Height	Number of Cubes
3	5	2	
4	6	5	
4	4	4	
7	8	3	
6	5.5	10	
14	3	6.5	
15	9	5	
8.7	13	5	

1. Write a formula to find the volume of a rectangular prism using the length, width, and height.

2. Write a rule to find the total **number of cubes** for each prism using the **area of the base** and the **height** of the prism.

3. A rectangular prism with dimensions 2, 5, and 8 can have three different bases, each with different dimensions.

   a. Find the area of each base.

   b. Use each base to find the volume of the rectangular prism.

4. Explain how the formula you wrote in Exercise 2 can be applied to determine the volume of a **cylinder**.

Activity 14: Pyramids and Cones

PURPOSE   Develop the relationship between the volume of a pyramid and its related prism and the relationship between a cone and its related cylinder.

MATERIALS   Scissors, tape, rice, and nets for the prism, pyramid, cylinder, and cone (pages 266–269).

GROUPING   Work individually or in pairs.

GETTING STARTED   Make copies of the nets for the prism, pyramid, cylinder, and cone on construction paper or tag board. Cut out each net and construct each model.

---

1. What is true about the areas of the bases of the prism and the pyramid?

2. What is true about the heights of the prism and the pyramid?

3. Estimate how many pyramids full of rice it will take to fill the prism.

Fill the pyramid with rice. Pour the rice into the prism. Repeat the process until the prism is full. (Be sure to record each time you pour rice into the prism.)

4. One prism full of rice = _____ pyramids full of rice.

5. Write a ratio that compares the volume of the pyramid to the volume of the prism.

6. On the basis of this exploration and the ratio you wrote in Exercise 5, write a rule to determine the volume of a pyramid based on the formula for the volume of a prism.

Repeat the experiment using the cylinder and the cone. Place the cylinder in a box or some container, since it is open at both ends.

1. Write a ratio that compares the volume of the cone to the volume of the cylinder.

2. On the basis of this exploration and the ratio you wrote in Exercise 1, write a rule to determine the volume of a cone based on the formula for the volume of a cylinder.

NET FOR PRISM

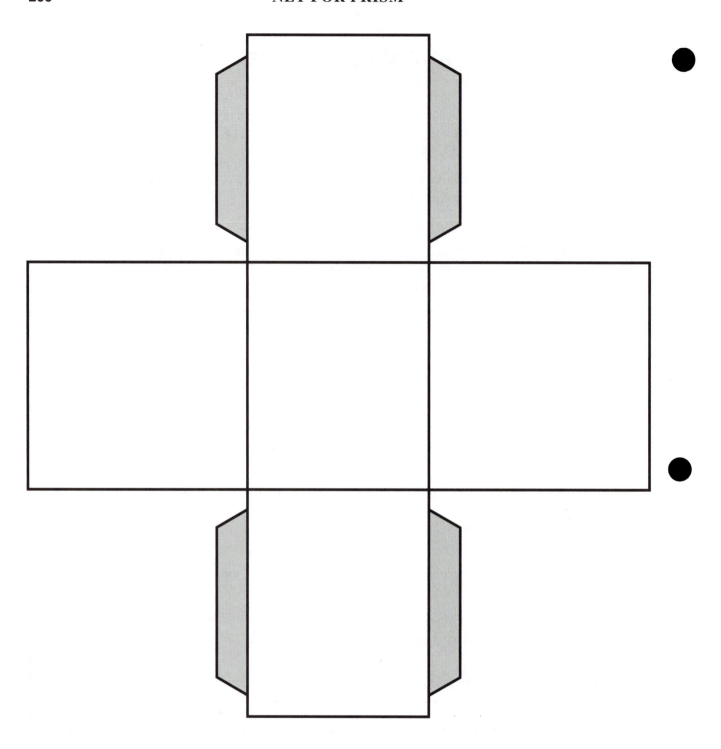

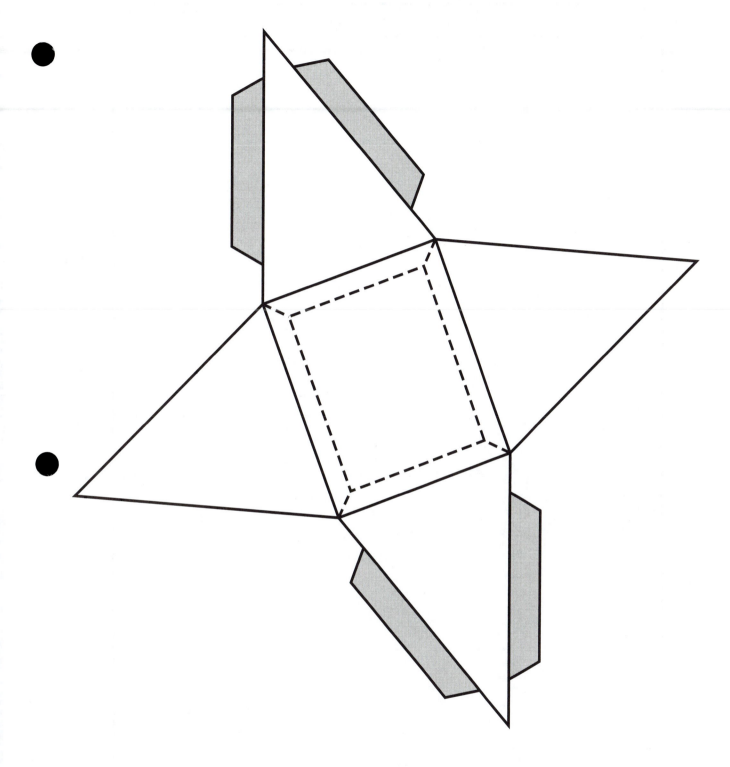

Cut on dotted line to open the pyramid.
Fold the tabs inside the pyramid.

NET FOR CYLINDER

Tab for overlap to tape

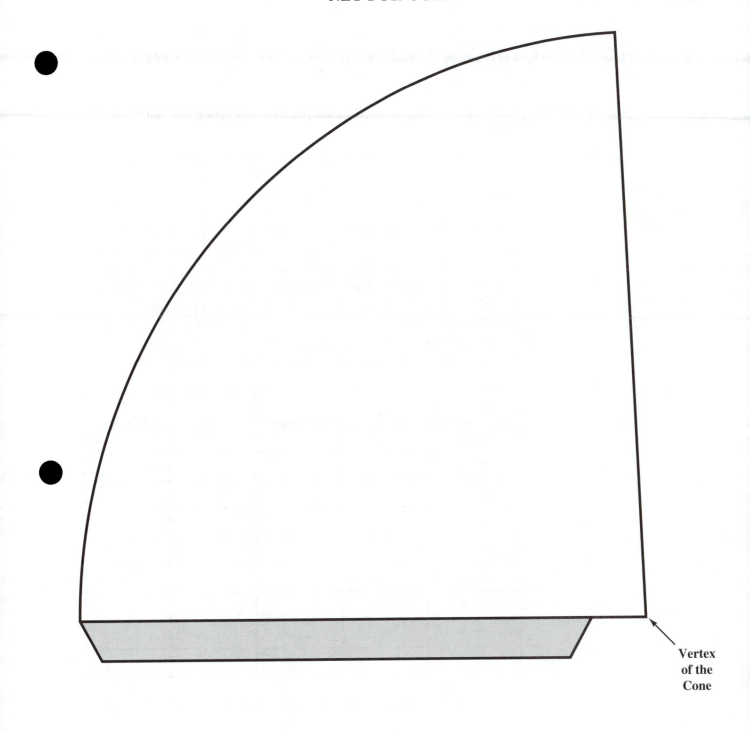

Vertex
of the
Cone

Activity 15: Compare Volume to Surface Area

PURPOSE   Develop an understanding of the relationship between volume and surface area.

MATERIALS   Scissors, tape, and several sheets of centimeter graph paper (page A–49)

GROUPING   Work individually or in pairs.

GETTING STARTED   Cut out a 17 cm × 24 centimeter rectangle from the graph paper. Cut one square from each corner of the paper. Fold up the sides and tape the edges to make an open-top box. Be careful not to overlap the paper. Use the squares on the paper to determine the area of the base, the altitude of the box, the surface area of the box, and the volume. Form new boxes by cutting 2 × 2, 3 × 3, etc., squares from each corner of a 17 cm × 24 cm rectangle cut from the graph paper. Find the new measurements and enter your data for each box in the table below.

Dimensions of Squares Cut Off	Area of Base	Altitude	Surface Area	Volume
1 × 1				
2 × 2				
3 × 3				
4 × 4				
5 × 5				
6 × 6				
7 × 7				
8 × 8				

1. What are the dimensions of the box with the greatest volume?

2. Does that box also have the greatest surface area?

3. As surface area increases, does the volume also increase? Explain.

EXTENSIONS   If you were a manufacturer of cereal boxes, how would you use the results of this activity to minimize costs and maximize profits?

Chapter 11 Summary

Perimeter, area, and volume are important concepts in geometry and have wide application in the real world. In fact, of all the topics studied in geometry, they are the source of more problems in everyday life than any other topic.

Research literature contains many references to the need for developmentally appropriate instruction about measurement and the concepts of perimeter, area, and volume. Students not only need to know *how* to measure, they must also know *what it is* that must be measured. Elementary students who learn formulas without a firm grasp of the basic concepts of measurement will have difficulty in their future study of geometry.

Activity 1 introduced units of measure and the need for Standard Units. Without international standardization of units, world trade would be impossible. This activity also reinforced the inverse relationship between the size of a unit of length and the number of the units used to measure a given length. That is, the smaller the unit (cm vs m), the larger number used to indicate the length of a given line segment (250 cm vs 2.5 m).

In Activity 2, perimeter was explored in a problem-solving setting where you discovered a pattern to determine the perimeter of any given number of regular polygons laid end to end. Using the meaning of perimeter—the distance around a figure—and finding the rule to determine it helped you develop a firm understanding of perimeter.

Activities 3 and 4 introduced area as covering with a given unit. Using a triangle and a square as units of area and determining the number of triangles or squares that would cover the different shapes helped develop spatial visualization and estimation skills.

Activities 6–8 developed the formulas needed to determine the areas of triangles and special quadrilaterals. Each new polygon was related to one previously studied. Through these activities, you found that all these area formulas have their foundation in the rectangular arrays used to illustrate multiplication.

In Activities 4 and 15, you investigated two concepts that are often misunderstood: the relationship between perimeter and area, and the relationship between surface area and volume. Most people believe that as the area or volume of an object increases so does its perimeter or surface area respectively. After completing these activities, you discovered that both of these assumptions are false.

Activities 9 and 10 introduced one of the most important theorems of geometry, the Pythagorean theorem. In Activity 9, you constructed a square on the hypotenuse of a right triangle using the pieces of the squares constructed on one of the other two sides. This firmly established the relationship between the area of the square on the hypotenuse and the sum of the areas of the squares on the other two sides.

You constructed triangles by putting squares together in Activity 10 to explore the converse of the Pythagorean theorem and to reinforce classification of triangles by angles and the Triangle Inequality theorem.

Activity 11 provided an opportunity to explore an apparent magic trick. The area of a square was divided into four parts and rearranged. In one case a unit of area disappeared; in the second, an additional unit of area was added. Conservation of area tells us that neither of these situations is possible. Through the application of coordinate geometry, the slope of a line or distance between points, you discovered that the eye was deceived when the parts of the square were rearranged.

Activities 12–15 developed the concept of volume and the formulas for the volumes of certain polyhedra. The process of finding a systematic method for counting cubes to determine volume was directly related to multiplying the dimensions of a rectangular array to count the squares when determining area.

By comparing a pyramid to a prism and a cone to a cylinder, you developed a connection among these solids and the formulas to find their volumes. This comparison strategy is the same that is used to find area formulas where related polygons were compared.

Throughout this chapter, you were involved in *doing mathematics*. You discovered patterns, made and tested conjectures, and then reasoned to develop formulas for perimeter, area, and volume. You also constructed new knowledge or reinforced prior knowledge about these important attributes. Your active involvement in developing the formulas helped enhance your conceptual understanding of these concepts.

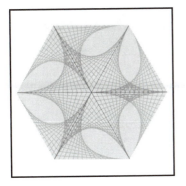

Chapter 12:
Motion Geometry and Tessellations

"Transformational geometry offers another lens through which to investigate and interpret geometric objects. To help them form images of shapes through different transformations, students can use physical objects, figures traced on tissue paper, mirrors or other reflective surfaces, figures drawn on graph paper, and dynamic geometry software. They should explore the characteristics of flips, turns, and slides and should investigate relationships among compositions of transformations. These experiences should help students develop a strong understanding of line and rotational symmetry, scaling, and properties of polygons."
—*Principles and Standards for School Mathematics*

In this chapter you will study the properties of a class of functions called isometries. An isometry is a mapping or transformation that preserves the distance between points. The transformations in this chapter include slides (translations), flips (reflections), turns (rotations), and glide reflections.

You will also investigate the properties of a dilation, a size transformation, that reduces or enlarges figures.

The chapter concludes with an exploration of symmetry—line symmetry, point symmetry, and rotational symmetry. Isometries, dilations, and symmetry are powerful new tools for studying congruence, similarity, and other geometric concepts.

Activity 1: Slides, Flips, and Turns

PURPOSE   Introduce translations, reflections, and rotations using pattern blocks.

MATERIALS   Pattern Blocks (pages A-7–A-11), a ruler, and a protractor

GROUPING   Work individually or in pairs.

SLIDES

Cover each shape below with pattern blocks. Slide the blocks the distance and direction indicated by the arrow and trace them.

1.

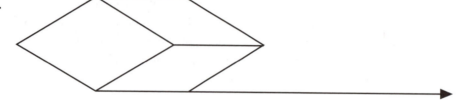

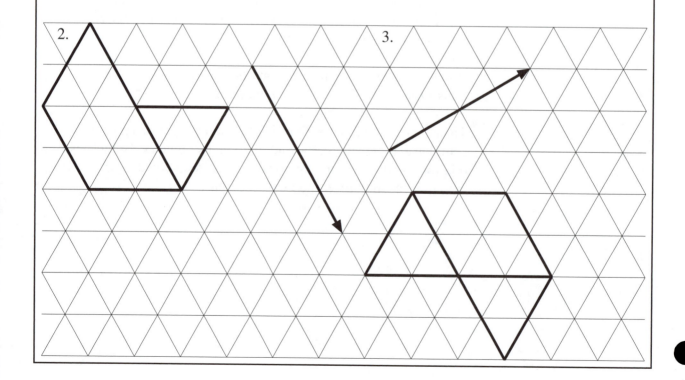

FLIPS

Cover each shape below with pattern blocks. Flip the blocks across the line and trace them.

1.

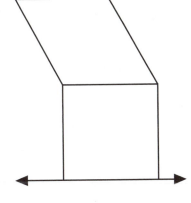

2.

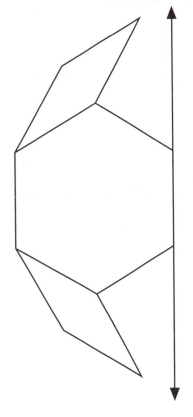

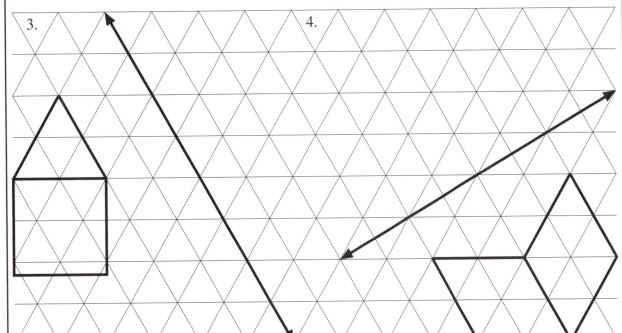

3.

4.

TURNS

Cover each shape with pattern blocks. Turn the blocks through the given angle in the indicated direction about point *O*. Record the result.

In the example at the right, the shaded block was turned 60° clockwise about point *O*.

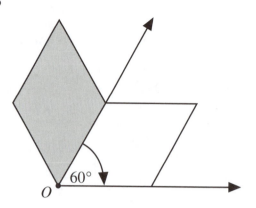

1. 60° counterclockwise

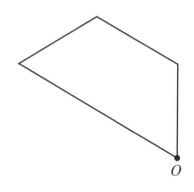

2. 90° clockwise

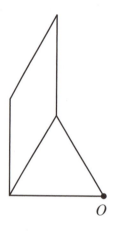

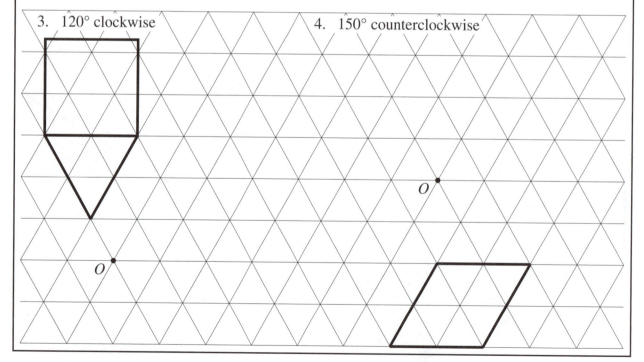

3. 120° clockwise

4. 150° counterclockwise

Activity 2: How Did You Do It?

PURPOSE	Identify slides and flips.
MATERIALS	Pattern Blocks (pages A-7–A-11)
GROUPING	Work individually or in pairs.
GETTING STARTED	Cover the shaded shapes with pattern blocks. Move the blocks to the next shapes by sliding or flipping them. Write the transformation that was used to make each move in the blank for each shape.

Example:

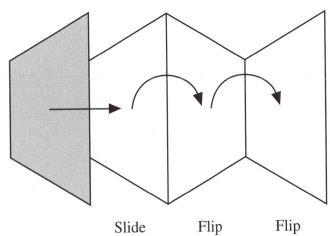

Slide   Flip   Flip

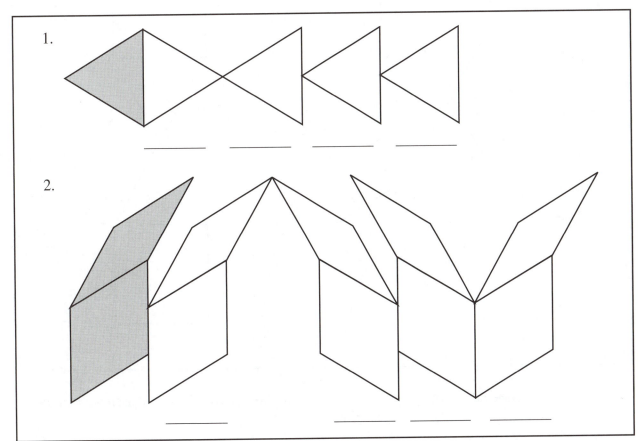

1.

2.

1.

2.

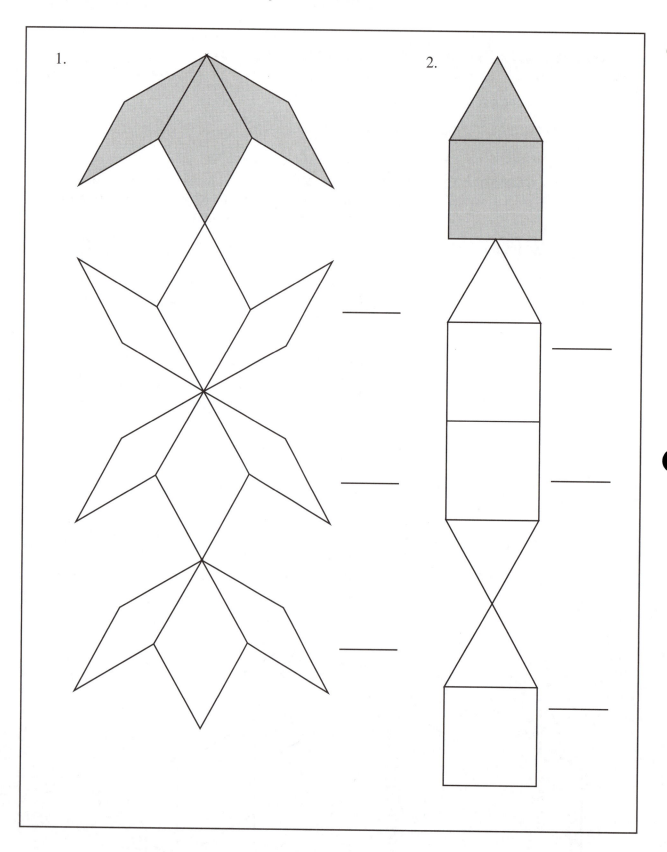

EXTENSIONS   Make up patterns similar to those above. Have a classmate identify the transformations.

Activity 3: Double Flips

PURPOSE   Identify transformations and investigate the composite of two reflections.

MATERIALS   Pattern Blocks (pages A-7–A-11), a ruler, and a protractor

GROUPING   Work individually or in pairs.

1. Cover the shape with a block. Flip the block across line $\overleftrightarrow{MA}$, then flip it across line $\overleftrightarrow{RK}$ and trace the block.

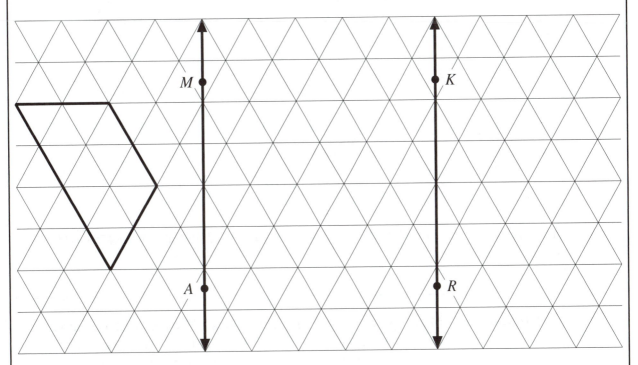

2. How is line $\overleftrightarrow{MA}$ related to line $\overleftrightarrow{RK}$?

3. What is the distance between the lines?

4. What single transformation has the same result as flipping the trapezoid across $\overleftrightarrow{MA}$ and then flipping it across $\overleftrightarrow{RK}$?

1. Cover the shape with a block. Flip the block across line $\overleftrightarrow{DO}$, then flip it across line $\overleftrightarrow{OT}$ and trace the block.

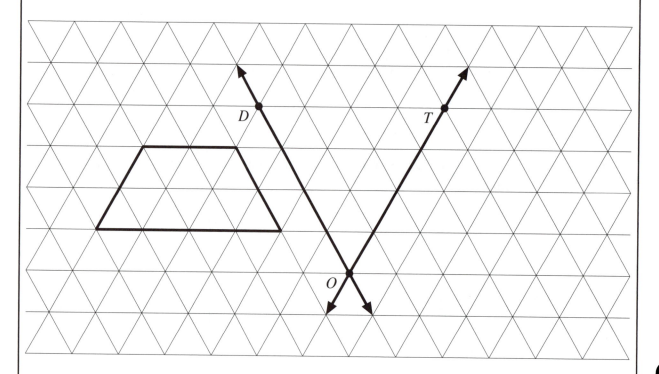

2. What is the measure of ∠DOT?

3. What single transformation has the same result as flipping the trapezoid across $\overleftrightarrow{DO}$, and then flipping it across $\overleftrightarrow{OT}$?

EXTENSIONS   Make up a design using pattern blocks. Use the design to make two problems like the previous ones in this activity, one where the design is flipped across two parallel lines and one where it is flipped across two intersecting lines. Have your partner perform the transformations, trace the results, and make the measurements as in the activities above.

Activity 4: Reflections

PURPOSE   Explore reflections and their properties.

MATERIALS   A compass, a centimeter ruler, a protractor, and a Mira

GROUPING   Work individually or in pairs.

GETTING STARTED   A Mira is a plastic drawing device that acts like a mirror. A Mira reflects objects, but since it is transparent, the image of an object reflected in it also appears behind the Mira.

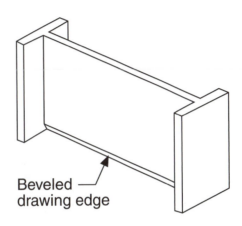

Beveled drawing edge

The drawing edge of a Mira is beveled. When using a Mira, place it with the beveled edge down. Look directly through the Mira from the side with the beveled edge to locate the image of the object behind the Mira.

Place your Mira so that the image of circle *A* fits on circle *B*. Hold the Mira steady with one hand and draw a line along the drawing edge.

Take away the Mira. The line you have drawn is called the *Mira line*. It represents the Mira. How does the Mira line appear to be related to points *A* and *B*?

For each pair of figures below, use a Mira to fit the image of one of the figures onto the other. Then draw the Mira line.

1.   2.   3.

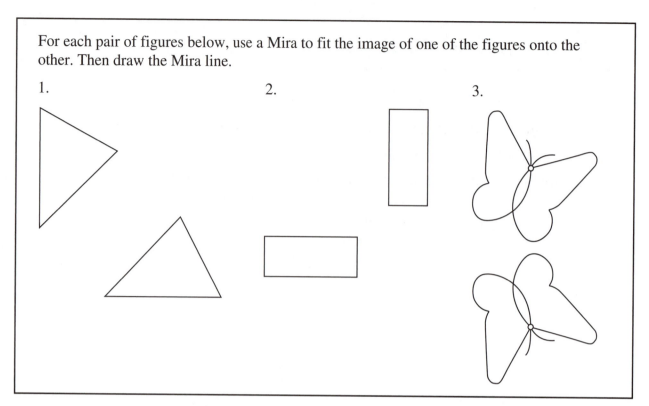

Use a Mira to draw the reflection of each figure through the given line.

1.   2.

3.   4.

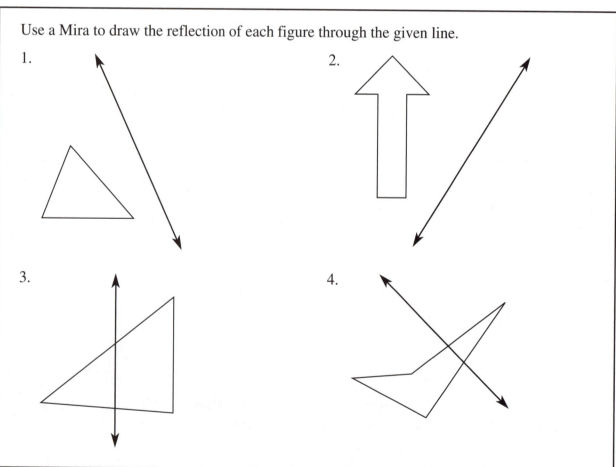

Use a Mira to mark the location of the reflection of each point through the line ℓ. Use prime notation to name each image point. For example, the image of point D would be named D′.

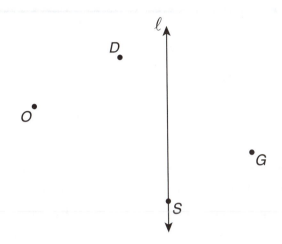

Draw line segments $\overline{DD'}$, $\overline{OO'}$, and $\overline{GG'}$. Label the points where line ℓ intersects these segments C, A, and T, respectively.

1. What is the relationship between the line ℓ and the segments $\overline{DD'}$, $\overline{OO'}$, and $\overline{GG'}$?

2. Where is point S located in relation to line ℓ?

3. What is the relationship between points S and S′?

Use a straightedge and compass to make the following constructions.

1. A line ℓ so that point P is the reflection of point A through ℓ.

2. A line ℓ so that pentagon R is the reflection of pentagon S through ℓ.

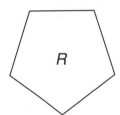

P •

• A

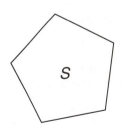

3. a. Explain how you constructed line ℓ.

   b. Why did you construct line ℓ as you did?

284 Chapter 12 • Motion Geometry and Tessellations

Use a straightedge and compass to construct the reflection of △ABC through line ℓ.

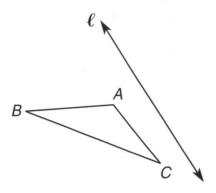

Use a Mira to reflect the figure *FLAG* through the line ℓ. Use prime notation to label the reflection.

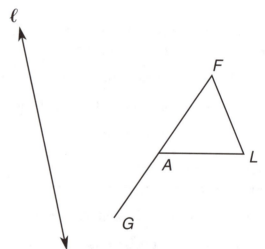

Measure the corresponding angles and segments in *FLAG* and its reflection image. Indicate the measures on the drawings.

1. Examine the measures of the angles. What can you conclude about the measure of an angle and the measure of its reflection?

2. Examine the lengths of the segments. What can you conclude about the length of a segment and the length of its reflection?

3. What can you conclude about a triangle and its reflection through a line?

4. a. Imagine tracing △FAL from F to A to L and back to F. What direction (clockwise or counterclockwise) would you move?

   b. Now imagine tracing the image △F′A′L′ from F′ to A′ to L′ and back to F′. What direction would you move?

   c. How does reflecting a figure through a line affect its orientation?

Copyright © 2004 Pearson Education, Inc. All rights reserved.

Activity 5: Translations

PURPOSE   Explore translations and their properties.

MATERIALS   A compass, a centimeter ruler, a piece of tracing paper, and a Mira

GROUPING   Work individually or in pairs.

1. Reflect $\triangle MAT$ through line ℓ_1. Use prime notation to label the reflection.

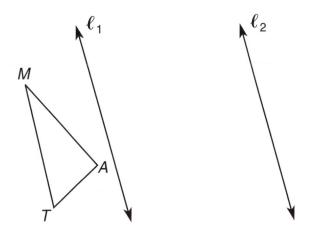

2. Reflect $\triangle M'A'T'$ through line ℓ_2. Let M'', A'', and T'' denote the images of M', A', and T', respectively.

3. Draw $\overline{MM''}$, $\overline{AA''}$, and $\overline{TT''}$.

4. Make a tracing of $\triangle MAT$. Slide it onto $\triangle M''A''T''$ by moving its vertices along the three "tracks" you have just drawn. Is it necessary to flip or to turn the tracing to do this?

5. What two relationships do the tracks appear to have?

This transformation is called a *translation*. The exercises illustrate the following definition:
A *translation* is the composite of two reflections through parallel lines.

1. Translate $\overline{ID}$ by reflecting it through line ℓ_3 and then reflecting the image through line ℓ_4. Let $\overline{ES}$ denote the final translation image of $\overline{ID}$ (E is the image of I and S is the image of D).

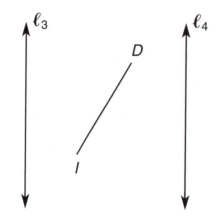

2. What is the distance between lines ℓ_3 and ℓ_4?

3. Measure IE and DS. How do these lengths compare with your answer in Exercise 2?

4. In what direction was $\overline{ID}$ translated?

5. a. Now translate $\overline{ID}$ by reflecting it through line ℓ_4 and then reflecting the image through line ℓ_3. Let $\overline{GL}$ denote the translation image of $\overline{ID}$ (G is the image of I and L is the image of D).

   b. Measure the lengths IG and DL. How do these lengths compare with the result in Exercise 2?

   c. In what direction was $\overline{ID}$ translated?

6. a. If a point is translated by reflecting it through two parallel lines that are x units apart, what is the distance between the point and its image?

   b. If a figure is translated by first reflecting it through line ℓ_5 and then reflecting its image through line ℓ_6, in what direction will the figure be translated?

   c. Use the translation of $\triangle MAT$ on page 285 to check your conclusions in Parts a and b.

Activity 6: Transforgeo

PURPOSE	Determine the images that are defined by a reflection or a translation.
MATERIALS	Two geoboards, six geobands, and dot paper (page A-50)
GROUPING	Work in pairs.
GETTING STARTED	*Transforgeo* is a game for two players. Each player needs a geoboard and three geobands.

GAME 1

- The players construct a reflecting line on their geoboards by connecting the pegs in the middle row with a geo-band.
- Using one geo-band, each player constructs a figure on the top half of a geoboard.
- When both players have completed their figures, they trade geoboards.
- The first player to construct the reflection of the figure on the geoboard is the winner. (The reflection must be approved by the builder of the original figure.)

Example:

Player 1 builds this figure on the top half of the geoboard.

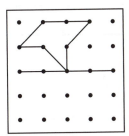

Player 2 must construct the reflection of the figure on the other half of the geoboard.

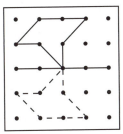

Play the game five times.

GAME 2

- The players construct a reflecting line on their geoboards by connecting the pegs on a diagonal with a geo-band.
- Using one geo-band, each player constructs a figure on half of a geoboard.
- When both players have completed their figures, they exchange geoboards.
- The first player to construct the reflection of the figure on the geoboard is the winner. (The reflection must be approved by the builder of the original figure.)

Example:

Player 1 builds this figure on half of the geoboard.

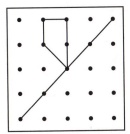

Player 2 must construct the reflection of the figure on the other half of the geoboard.

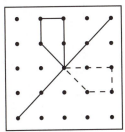

Play the game five times.

GAME 3

- Each player constructs a design on a geoboard using one geoband.
- With the second geoband, indicate the direction and distance the figure is to be translated. (The translation must not move the figure off the geoboard.)
- When both players have completed their figures, they exchange geoboards.
- The first player to construct the translation of the figure on the geoboard is the winner. (The translation of the figure must be approved by the builder of the original figure.)

Example:

Player 1 builds this figure on the geoboard. (Direction and distance of translation.)

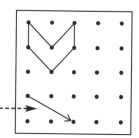

Player 2 must construct the translation of the figure on the geoboard.

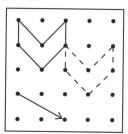

Play the game five times.

1. Reflect each figure through the given line.

   a. 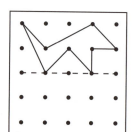   b. 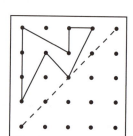   c.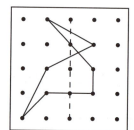

2. Translate each figure the distance and direction indicated by the arrow.

   a. 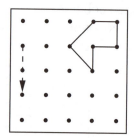   b.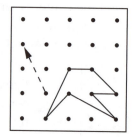

EXTENSIONS Play the games on dot paper using 10×10 squares to represent the geoboard.

Activity 7: Rotations

PURPOSE   Explore rotations and their properties.

MATERIALS   A compass, a centimeter ruler, a protractor, tracing paper, a straight pin or tack, and a Mira

GROUPING   Work individually or in pairs.

GETTING STARTED   The two drawings in Figure 1 are identical, but the duck is neither a reflection nor a translation of the rabbit. Use a Mira to verify that the duck is not a reflection of the rabbit and a tracing of the duck to verify that it is not a translation.

Place a piece of tracing paper over the rabbit. Pin it at point *P*. Trace the rabbit and then turn the paper about the pin until the rabbit coincides with the duck. This illustrates why the duck is called a *rotation image* of the rabbit.

P.

Figure 1

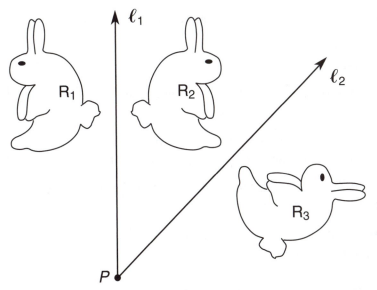

The drawings in Figure 2 show that if the rabbit, R₁ is reflected through line ℓ₁ and its image R₂ is reflected through line ℓ₂, then the result is the duck, R₃. Verify this with your Mira.

This example shows that the rotation that transforms the rabbit into the duck is the composite of two reflections through intersecting lines. This illustrates the following definitions:

A *rotation* is the composite of two reflections through intersecting lines.

The *center of rotation* is the point at which the two lines intersect.

Figure 2

1. What is the measure of the acute angle formed by lines ℓ_1 and ℓ_2?

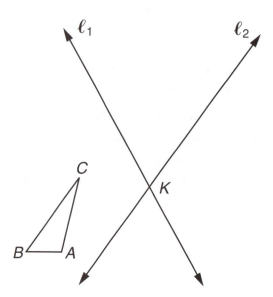

2. Rotate $\triangle ABC$ by reflecting it through line ℓ_1 and then reflecting its image through line ℓ_2. Let X, Y, and Z be the rotation images of points A, B, and C, respectively.

   a. Draw and measure $\angle BKY$. (Note that the vertex of $\angle BKY$ is at the center of rotation and that one side of the angle contains a point B on the original figure. The other side contains the rotation image Y of point B. The measure of an angle formed this way is called the *magnitude of the rotation*.)

   b. How does the magnitude of the rotation compare with your answer to Exercise 1?

   c. In what direction, clockwise or counterclockwise, was $\triangle ABC$ rotated?

3. Rotate $\triangle ABC$ by reflecting it through line ℓ_2 and then reflecting its image through line ℓ_1. Let R, S, and T be the rotation images of points A, B, and C, respectively.

   a. What is the magnitude of the rotation? How does it compare with your answer to Exercise 1?

   b. In what direction was $\triangle ABC$ rotated?

4. a. On the figure in Exercise 1, draw three circles with center K and radii $\overline{BK}$, $\overline{CK}$, and $\overline{AK}$, respectively. What do you notice about the circles?

   b. Make a tracing of $\triangle ABC$. Slide it onto $\triangle XYZ$ by moving its vertices along the circular "tracks" you have just drawn. Is it necessary to flip or to turn the tracing?

5. a. A point is rotated by reflecting it through two intersecting lines that form an acute angle with measure $x°$. What will be the magnitude of the rotation?

   b. If a figure is rotated by first reflecting it through line ℓ_3 and then reflecting its image through line ℓ_4, in what direction will the figure be rotated?

Activity 8: Glide Reflections

PURPOSE Explore glide reflections and their properties.

MATERIALS A compass, a centimeter ruler, a protractor, and a Mira

GROUPING Work individually or in pairs.

GETTING STARTED In Figure 1, footprint F_3 is a glide reflection of footprint F_1. As the name suggests, a glide reflection is a glide (or translation) followed by a reflection. However, the reflecting line must be parallel to the direction of the glide.

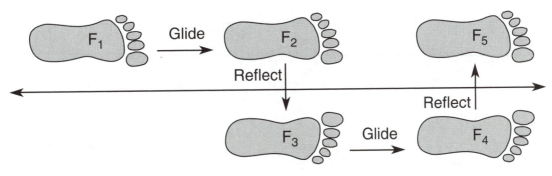

Figure 1

A glide reflection maps $\triangle ABC$ onto $\triangle A'B'C'$. Draw segments $\overline{AA'}$, $\overline{BB'}$, and $\overline{CC'}$ and find their midpoints. What appears to be true about the midpoints of the segments?

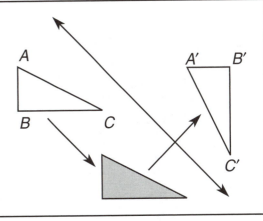

A glide reflection maps $\triangle XYZ$ onto $\triangle X'Y'Z'$. Find the reflecting line and draw the glide image of $\triangle XYZ$.

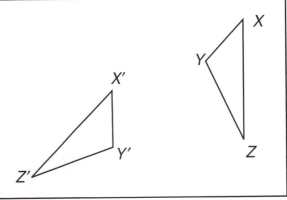

Study Figure 1. What single transformation is equivalent to the result of using a glide reflection twice?

Identify the transformation that will map the shaded hawk in Figure 2 onto each lettered hawk.

1. Hawk A _____   2. Hawk B _____

3. Hawk C _____   4. Hawk D _____

5. Hawk E _____   6. Hawk F _____

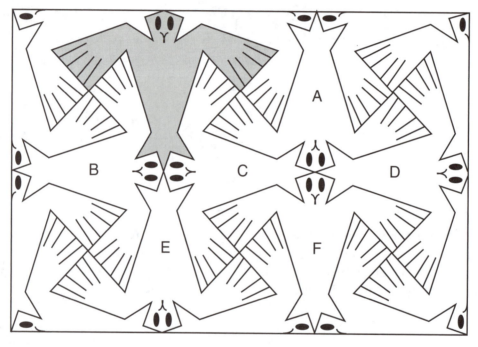

Figure 2

EXTENSIONS   Complete the following statements:

1. The composite of two reflections through parallel lines is a _____.

2. The composite of two reflections through intersecting lines is a _____.

3. The composite of three reflections through parallel lines is a _____.

4. The composite of three reflections through concurrent lines is a _____.

5. The composite of three reflections through lines that are not all parallel and not all concurrent is a _____.

Activity 9: Size Transformations

PURPOSE   Explore size transformations (dilations) and their properties.

MATERIALS   A centimeter ruler and a protractor

GROUPING   Work individually or in pairs.

GETTING STARTED   A *dilation* is a transformation that is defined by a point, the *center of dilation*, and a *scale factor*. In the figure below:

- Draw rays $\overrightarrow{PA}$, $\overrightarrow{PB}$, and $\overrightarrow{PC}$.

- Mark point A' on $\overrightarrow{PA}$ such that A is between P and A' and $PA = AA'$.

- Mark point B' on $\overrightarrow{PB}$ such that B is between P and B' and $PB = BB'$.

- Mark point C' on $\overrightarrow{PC}$ such that C is between P and C' and $PC = CC'$.

- Draw $\triangle A'B'C'$. $\triangle A'B'C'$ is the *dilation image* of $\triangle ABC$.

$Q.$

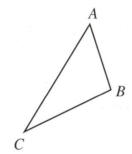

$P.$

1. Measure the sides and angles of $\triangle ABC$ and $\triangle A'B'C'$. Record the measures in the table.

$\triangle ABC$	
$m\angle A$	
$m\angle B$	
$m\angle C$	
AB	
AC	
BC	

$\triangle A'B'C'$	
$m\angle A'$	
$m\angle B'$	
$m\angle C'$	
$A'B'$	
$A'C'$	
$B'C'$	

2. Are $\triangle ABC$ and $\triangle A'B'C'$ similar? Explain.

3. Find the following ratios. What do you notice?

$$\frac{A'B'}{AB} =$$

$$\frac{A'C'}{AC} =$$

$$\frac{B'C'}{BC} =$$

4. a. What is the ratio of PA' to PA?

    b. How does this ratio compare to the ones in Exercise 3?

    c. Point P is the center of the dilation and the ratio $PA' : PA$ is the scale factor. Why does it make sense to call this ratio a scale factor?

5. a. Find the midpoints of segments $A'B'$, $B'C'$, and $C'A'$, and label them R, S, and T, respectively. Draw $\triangle RST$.

    b. Drawing $\triangle RST$ divided $\triangle A'B'C'$ into four triangles. How do these triangles appear to be related?

    c. How is $\triangle RST$ related to $\triangle ABC$?

    d. How is the area of $\triangle A'B'C'$ related to the area of $\triangle ABC$?

6. In the figure at the beginning of the activity:

    • Mark point A'' on $\overrightarrow{PA}$ such that A is between P and A'' and $AA'' = 2 \cdot PA$.

    • Mark point B'' on $\overrightarrow{PB}$ such that B is between P and B'' and $BB'' = 2 \cdot PB$.

    • Mark point C'' on $\overrightarrow{PC}$ such that C is between P and C'' and $CC'' = 2 \cdot PC$.

    • Draw $\triangle A''B''C''$.

7. a. Find the ratio of the length of each side of $\triangle A''B''C''$ to the length of the corresponding side in $\triangle ABC$.

   b. Is $\triangle ABC$ similar to $\triangle A''B''C''$? Explain.

   c. What is the scale factor of the dilation (the ratio of PA'' to PA)? How does it compare to the ratios in Part (a)?

   d. Trisect the sides of $\triangle A''B''C''$. Show how you could connect the trisection points to show that the area of $\triangle A''B''C''$ is nine times the area of $\triangle ABC$.

8. a. Is $\triangle A''B''C''$ similar to $\triangle A'B'C'$?

   b. What is the scale factor of the dilation that maps $\triangle A'B'C'$ to $\triangle A''B''C''$ (the ratio of PA'' to PA')?

   c. How is the area of $\triangle A''B''C''$ related to the area of $\triangle A'B'C'$?

9. In the figure at the beginning of the activity:

   • Draw rays $\overrightarrow{QA}$, $\overrightarrow{QB}$, and $\overrightarrow{QC}$.

   • Mark point X on $\overrightarrow{QA}$ such that A is between Q and X and $QA = AX$.

   • Mark point Y on $\overrightarrow{QB}$ such that B is between Q and Y and $QB = BY$.

   • Mark point Z on $\overrightarrow{QC}$ such that C is between Q and Z and $QC = CZ$.

   • Draw $\triangle XYZ$.

10. What effect does moving the center of dilation have on the result of the transformation?

11. A dilation maps $\triangle JKL$ to $\triangle J'K'L'$. If the center of dilation is M:

    a. What is the scale factor of the dilation?

    b. How are the lengths of the sides of the two triangles related?

    c. How are the areas of the two triangles related?

Activity 10: Coordinate Transformations

PURPOSE   Investigate the effects of coordinate transformations on graphs.

MATERIALS   Graph paper, tracing paper, a Mira, a ruler, and a protractor

GROUPING   Work individually or in pairs.

1. Plot and label the points $A(1, 3)$, $B(2, -3)$, and $C(-2, 1)$ on a pair of coordinate axes. Draw $\triangle ABC$.

2. Plot and label the points $A'(6, 3)$, $B'(7, -3)$, and $C'(3, 1)$ on the same pair of coordinate axes. Draw $\triangle A'B'C'$.

3. How are the coordinates of A', B', and C' related to the coordinates of A, B, and C?

4. What transformation maps $\triangle ABC$ onto $\triangle A'B'C'$?

5. a. The coordinates of points A'', B'', and C'' are formed from the coordinates of points A, B, and C by adding 3 to the respective y-coordinates. Without plotting points, predict the transformation that will map $\triangle ABC$ onto $\triangle A''B''C''$.

    b. Check your prediction in Part a by finding the coordinates of points A'', B'', and C'' and plotting the points.

6. a. $\triangle XYZ$ is the translation image when $\triangle ABC$ is translated 4 units to the left and 2 units down. Without performing the translation, how can you determine the coordinates of points X, Y, and Z from the coordinates of points A, B, and C?

    b. Check your answer to Part a by finding the coordinates of points X, Y, and Z and plotting the points.

---

1. Plot and label the points $P(1, 6)$, $Q(3, 5)$, and $R(4, 2)$ on a pair of coordinate axes. Draw $\triangle PQR$.

2. a. The coordinates of points P'', Q'', and R'' are formed from the coordinates of P, Q, and R by multiplying the respective x-coordinates by -1. Find the coordinates of P'', Q'', and R''.

    b. Plot and label the points P'', Q'', and R''. Draw $\triangle P''Q''R''$. What transformation maps $\triangle PQR$ onto $\triangle P''Q''R''$?

3. a. What would you do to the coordinates of points P, Q, and R to reflect $\triangle PQR$ through the x-axis?

    b. Check your answer to Part (a) by reflecting $\triangle PQR$ through the x-axis and finding the coordinates of the images of points P, Q, and R.

1. a. The vertices of a triangle are $S(2, 6)$, $T(4, 3)$, and $U(1, 4)$, and the points $S'(-2, -6)$, $T'(-4, -3)$, and $U'(-1, -4)$ are their images under a transformation. Without plotting points, predict what transformation maps $\triangle STU$ onto $\triangle S'T'U'$. **HINT:** This is the composite of two transformations.

   b. Check your prediction by plotting the points.

2. a. $\triangle STU$ is mapped onto $\triangle S''T''U''$ by reflecting $\triangle STU$ through the y-axis and then translating the image down 3 units (a glide reflection). Without performing the glide reflection, find the coordinates of points S'', T'', and U''.

   b. Check your answer by performing the glide reflection and finding the coordinates of S'', T'', and U''.

---

1. Plot and label the points $X(4, 1)$, $Y(4, 3)$, and $Z(7, 1)$ on a pair of coordinate axes. Draw $\triangle XYZ$.

2. Plot and label the points $X'(1, 4)$, $Y'(3, 4)$, and $Z'(1, 7)$ on the same pair of coordinate axes. Draw $\triangle X'Y'Z'$.

3. How are the coordinates of X', Y', and Z' related to the coordinates of X, Y, and Z?

4. What transformation maps $\triangle XYZ$ onto $\triangle X'Y'Z'$?

5. Plot and label the points $X''(-1, -4)$, $Y''(-3, -4)$, and $Z''(-1, -7)$ on the same pair of coordinate axes. Draw $\triangle X''Y''Z''$.

6. How are the coordinates of points X'', Y'', and Z'' related to the coordinates of X, Y, and Z?

7. What transformation maps $\triangle XYZ$ onto $\triangle X''Y''Z''$?

---

1. What single transformation has the same effect as reflecting a triangle through the line $y = x$ and then reflecting the image through the y-axis?

2. If the coordinates of the vertices of the triangle are (a, b), (c, d), and (e, f), what are the coordinates of the images of the vertices under this transformation?

3. Check your answers to Exercises 1 and 2 by drawing a triangle and using a Mira to perform the two reflections.

EXTENSIONS

1. Write a summary of the results of this activity.

2. Investigate the effects on a triangle when the coordinates of its vertices are multiplied by positive constants.

Activity 11: Cut It Out

PURPOSE	Introduce line, rotational, and point symmetry.
MATERIALS	Paper, scissors, ruler, protractor, and a straight pin or tack
GROUPING	Work individually or in pairs.
GETTING STARTED	Draw the angle with the given measure, fold the paper along the lines as shown, and cut out the shapes. Do the operations carefully, as accuracy is critical.

LINE SYMMETRY

1. On a separate sheet of paper, draw two lines, $\overleftrightarrow{AB}$ and $\overleftrightarrow{BC}$, that intersect at point B and such that $m \angle ABC = 60°$.

2. Draw an irregular polygon that has point B in its interior.

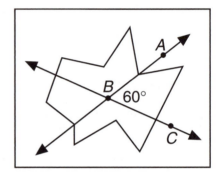

3. Carefully fold the paper along line $\overleftrightarrow{AB}$ so that the figure is on the outside of the paper and cut along the edges of the polygon.

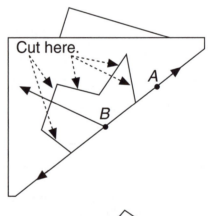

4. Turn the paper over and cut along any remaining edges of the polygon.

5. Open the paper. What do you observe?

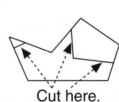

A geometric figure that can be folded along a line in such a way that the two halves are congruent and coincide is said to have *line symmetry*. The line along the fold is the *line of symmetry*.

6. What is the line of symmetry of the figure?

ROTATIONAL SYMMETRY

1. Fold the figure along the other line, $\overleftrightarrow{BC}$. **NOTE:** This is the second fold.

2. Cut away any portion of the figure where there is only one layer of paper. Turn the paper over and cut off the undoubled portion.

3. Open the paper. How many lines of symmetry does the figure have?

4. Continue folding the paper along lines $\overleftrightarrow{AB}$ and $\overleftrightarrow{BC}$ and cutting off the undoubled paper until there are no undoubled parts. If you have measured, folded, and cut accurately, this should require just one more fold (a total of three folds altogether).

5. Unfold the final figure. How many lines of symmetry does it have?

6. Flatten the figure on a piece of paper and trace its outline. Stick a pin through point B and rotate the figure about the pin. What do you observe?

A figure has rotational *symmetry* if rotating it through an angle less than 360° about some point makes the figure coincide with itself. The *angle of rotational symmetry* is the measure of the angle through which the figure was rotated.

7. What are the angles of rotational symmetry for the figure?

POINT SYMMETRY

1. On a separate sheet of paper, draw two lines, $\overleftrightarrow{AB}$ and $\overleftrightarrow{BC}$, that intersect at point B and such that $m \angle ABC = 90°$. Draw an irregular polygon that has point B in its interior.

2. Carefully fold the paper along line $\overleftrightarrow{AB}$ and cut along the edge of the polygon. Turn the paper over and cut along any of the remaining edges of the polygon. Continue the process of folding on lines $\overleftrightarrow{AB}$ and $\overleftrightarrow{BC}$ and cutting off the undoubled paper until there are no undoubled parts.

3. Unfold the final figure. How many lines of symmetry does it have?

4. What is the angle(s) of rotational symmetry for the figure?

Any figure that has 180° rotational symmetry is said to have *point symmetry* about the point of rotation.

5. Draw five segments that pass through point B and have their endpoints on the edge of the figure. How is point B related to the segments?

1. Repeat the experiment for two lines $\overleftrightarrow{AB}$ and $\overleftrightarrow{BC}$ that intersect to form a 45° angle. In the table below, record the maximum number of folds that were made before there was no excess paper to cut off.

Measure of ∠ABC	90°	60°	45°		30°		20°	18°	10°
Maximum Number of Folds	2	3	4	5					
Number of Lines of Symmetry	2	3	4			8			

2. a. Look for a pattern to find the missing entries in the table.

   b. Describe the pattern.

   c. Check your prediction for a 30° angle.

REGULAR POLYGONS

An *n*-gon is a polygon with *n* sides.

1. How many lines of symmetry does a regular *n*-gon have?

2. What are the rotational symmetries of a regular *n*-gon?

3. Which regular *n*-gons have point symmetry?

EXTENSIONS

1. a. Use the pattern in the table above. According to the pattern, if you start with two lines intersecting at an 80° angle, how many lines of symmetry should the resulting figure have?

   b. Is this possible?

   c. What do you think would happen if you started with an 80° angle?

   d. Try the experiment starting with an 80° angle. (**HINT:** Use a large piece of paper.) What happened? Why?

2. If a figure has exactly two lines of symmetry, what must be true about the lines of symmetry? Explain.

Activity 12: Draw It

PURPOSE	Apply the concept of symmetry to drawing figures.
MATERIALS	Graph paper and either a mirror or a Mira
GROUPING	Work individually.

1. Use the grid to complete the drawing of the figure at the right. The final figure should be symmetric with respect to the vertical line.

2. Place a mirror or a Mira on the line so that you can see the reflection of the left half of the picture in it. What do you observe?

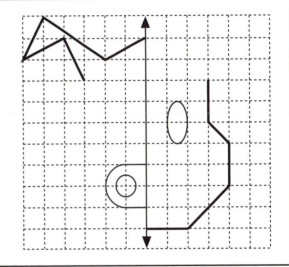

1. Use the grid to complete the drawing of the figure at the right. The final figure should be symmetric with respect to the point *O*.

2. Rotate the page 180° and look at the figure. What do you observe?

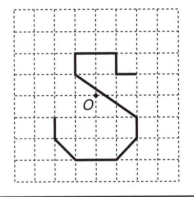

EXTENSIONS

1. Draw a figure that has line symmetry but no rotational symmetries.

2. Draw a figure that has point symmetry but no line symmetries.

3. Draw a figure that has rotational symmetry but neither point nor line symmetries.

4. Make up a design that has one or more lines of symmetry. Draw part of it on graph paper and have a classmate complete the design using symmetries as in the activity above.

Activity 13: Tessellations

PURPOSE Construct tilings using regular polygons and explore regular and semi-regular tessellations.

MATERIALS Scissors, tape, a ruler, five index cards, and three sets of Regular Polygons (page A-51) for each group of students (Copying the sets on card stock works well.)

GROUPING Work individually or in groups of 2–3.

GETTING STARTED A *tessellation* is a tiling that uses congruent figures to cover the plane without any gaps or overlaps. The term tessellation comes from *tessella*, the Latin word for the small square tiles used in ancient Roman mosaics.

There are many interesting questions related to tessellations. For example: **Which regular polygons will tessellate?** To find out, complete the following experiment.

- Carefully cut out the regular polygons and sort them by shape.
- Mark a point *P* near the center of a separate sheet of paper.
- Start with the equilateral triangles. Place a vertex of one of the triangles on *P*. Continue placing triangles edge-to-edge to cover the region around *P*. Record your results in the table.
- Continue this process until you have tested each regular polygon.

Polygon	Measure of Each Interior Angle	Number of Angles at *P*	Sum of the Measures of the Angles at *P*	Does the Polygon Tessellate?
Triangle				
Square				
Pentagon				
Hexagon				
Heptagon				
Octagon				
Nonagon				
Decagon				
Dodecagon				

1. What must be true about the measure of an interior angle of a regular polygon in order for the polygon to tessellate?

2. a. What are the only regular polygons that will tessellate?

   b. Make a sketch to show that each polygon in Part a will tile the plane. These are the *regular* tessellations.

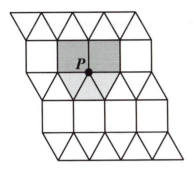

3. Use your result from Exercise 1 to show that the polygons in Exercise 2 are the only ones that will tessellate.

In the experiment, you discovered how one regular polygon can be used to create a tessellation. Tessellations can also be created using combinations of regular polygons. For example, the tessellation at the left was created using two squares and three equilateral triangles.

4. Find a different way to arrange two squares and three equilateral triangles around a point *P* to create a tessellation. Make a sketch of the tessellation.

5. Create four more tessellations that include more than one type of regular polygon. Make a sketch of each one.

A tessellation that uses a combination of regular polygons joined edge-to-edge and has the same arrangement of polygons around every vertex is *semi-regular*. The tessellation above is semi-regular, but the one below is not since the arrangements of polygons at vertices *Q* and *R* are different.

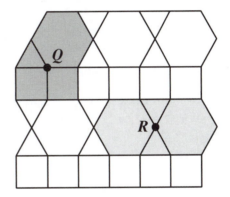

6. Which of the tessellations you created in Exercises 4 and 5 are semi-regular? If a tessellation isn't semi-regular, explain why not.

Now let's explore tessellating the plane with irregular polygons.

7. a. Use a ruler to draw a non-equilateral triangle on an index card and cut it out.

   b. Trace around your triangle to create a tessellation. Be sure corresponding sides of the triangles are joined edge-to-edge.

   c. Compare your triangle and tessellation with those of several classmates. Were there any triangles that would not tessellate?

   d. Will any triangle tessellate? If so, show how to arrange the triangles around a point to create a tessellation.

8. a. Draw a convex quadrilateral on an index card and cut it out.

   b. Trace around the quadrilateral to create a tessellation. Be sure to join the quadrilaterals edge-to-edge.

   c. Do you think any convex quadrilateral will tessellate? If so, show how to arrange the quadrilaterals around a point to create a tessellation. If not, explain why not.

9. Repeat Exercise 8 using a concave quadrilateral.

EXTENSIONS Follow the steps below to learn how translations and rotations can be used to create tessellations.

Using Translations:

Step 1: Draw a parallelogram on an index card and cut it out.

Step 2: Draw a shape along one side of the parallelogram. Cut it out and translate it to the opposite side. Tape on the translated piece.

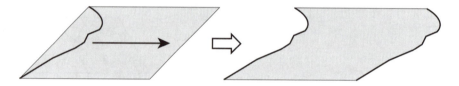

Step 3: Repeat Step 2 on the other two parallel sides.

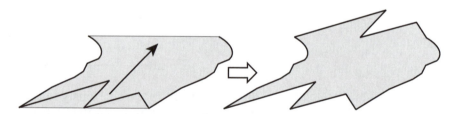

Step 4: Use your shape to create a tessellation.

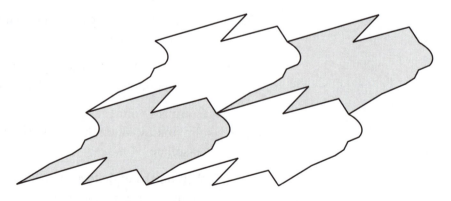

Using Rotations:

Step 1: Draw an equilateral triangle on an index card and cut it out.

Step 2: Draw a shape along one side of the triangle. Cut it out and rotate it 60° about an endpoint to modify a second side. Tape on the rotated piece.

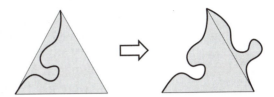

Step 3: Find the midpoint of the remaining side. Draw a shape along one half of the side. Cut out the shape, rotate it 180° about the midpoint, and tape it to the other half of the side.

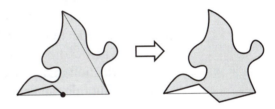

Step 4: Use your shape to create a tessellation.

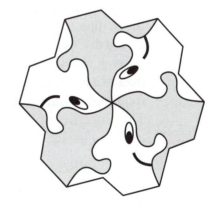

Chapter 12 Summary

In Activities 1–8, you studied four isometries—translations, reflections, rotations, and glide reflections—and discovered some of their properties. The first three transformations were introduced in Activities 1 and 2 through the use of pattern blocks and the less formal terminology of slides, flips, and turns. Because these terms are simple and very descriptive, they are the ones most often used with elementary students. Formal terminology and a variety of tools—tracing paper, Miras, geoboards, rulers, compasses, and protractors—were used in the later activities to perform the transformations and to study their properties.

Activity 3 introduced the idea that slides and turns result from performing two successive flips. This idea was explored in greater detail in Activities 5 and 7. Activity 4 extended the idea of a flip to the more general concept of a *reflection* or mirror image. A fundamental property of reflections was developed in the activity: the reflecting line is the perpendicular bisector of the segment connecting a point and its reflection image. This property makes it possible to reflect objects using ruler and compass constructions or coordinate methods.

In Activity 5, you discovered that a *translation* is the composite of two reflections through parallel lines. The distance the object is translated is twice the distance between the lines and the translation is in the direction from the first reflecting line to the second. In Activity 7, you discovered that a *rotation* is the composite of two reflections through intersecting lines. The magnitude of the rotation is twice the measure of the acute angle formed by the reflecting lines and the rotation is in the direction from the first reflecting line to the second. The properties of the only other isometry, a *glide reflection*, were explored in Activity 8.

You also discovered that in addition to preserving the distance between points, all four transformations preserved the measure of angles, collinearity of points, and congruence. The orientation of objects was preserved by translations and rotations, but reversed by reflections and glide reflections.

In Activity 9, you explored similarity transformations, or *dilations*, and saw how they enlarge or reduce a figure. A dilation is defined by a point, the center of dilation, and a scale factor. Dilations do not preserve the distance between points, however, they do preserve collinearity of points and the measures of angles. A figure and its dilation image are similar and, if the scale factor is k, the area of the image is k^2 times the area of the original figure.

In Activity 10, you used coordinate techniques to study the relationship between the geometric transformations and algebraic transformations such as adding or subtracting a constant and multiplying or dividing by a constant.

Activities 11 and 12 introduced the concept of symmetry. There are three kinds of symmetry. If an object has *line symmetry*, it can be divided into two halves in such a way that each half is the mirror image of the other.

An object has *rotational symmetry* if it can be made to coincide with itself by a rotation of less than 360° about a point. Objects that have a rotational symmetry of 180° are said to have *point symmetry*. If an object has point symmetry, it appears the same when viewed right-side-up or up-side-down.

Line and rotational symmetry are very common in nature. The body structure of most animals, including humans, possesses line symmetry. Some trees and the leaves of many plants also exhibit line symmetry, while many flowers exhibit rotational symmetries.

In the last activity, you used geometric transformations to study tessellations, tilings that use congruent shapes to completely cover the plane without any gaps or overlaps. You discovered that only three regular polygons—equilateral triangles, squares, and hexagons—tessellate and that all triangles and quadrilaterals tessellate.

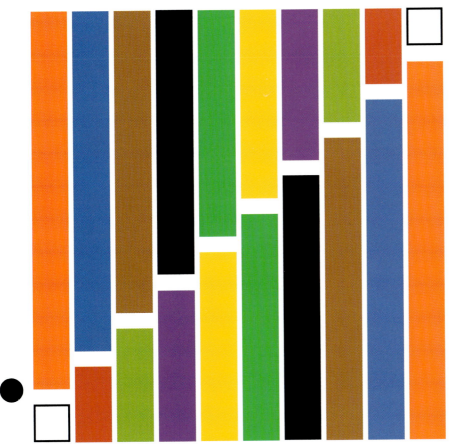

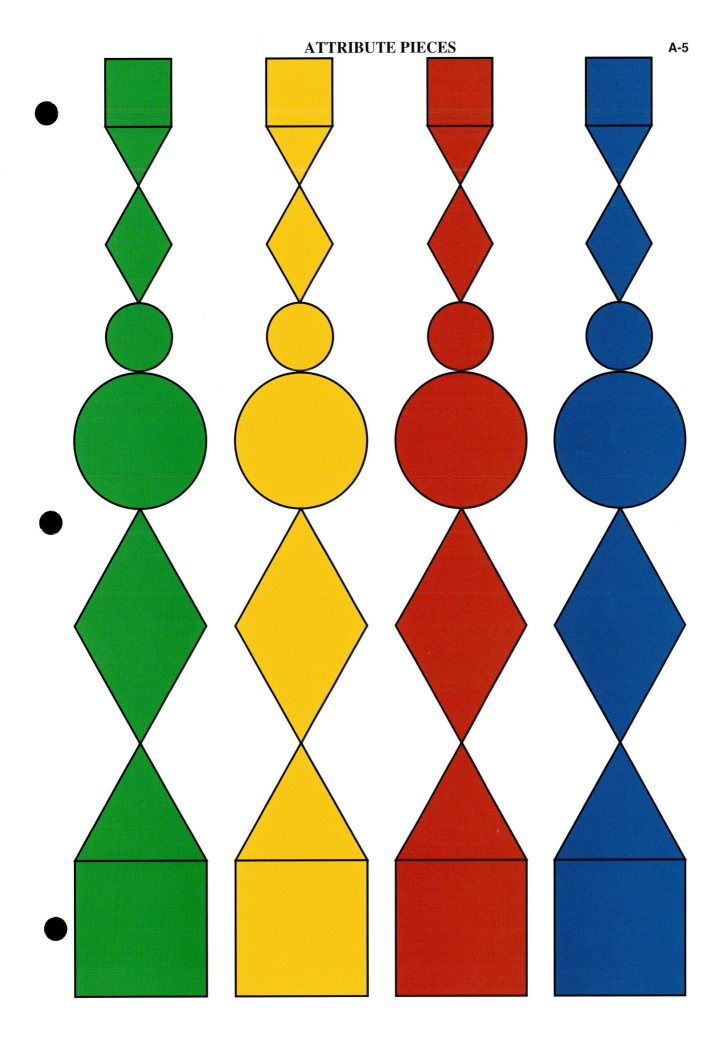

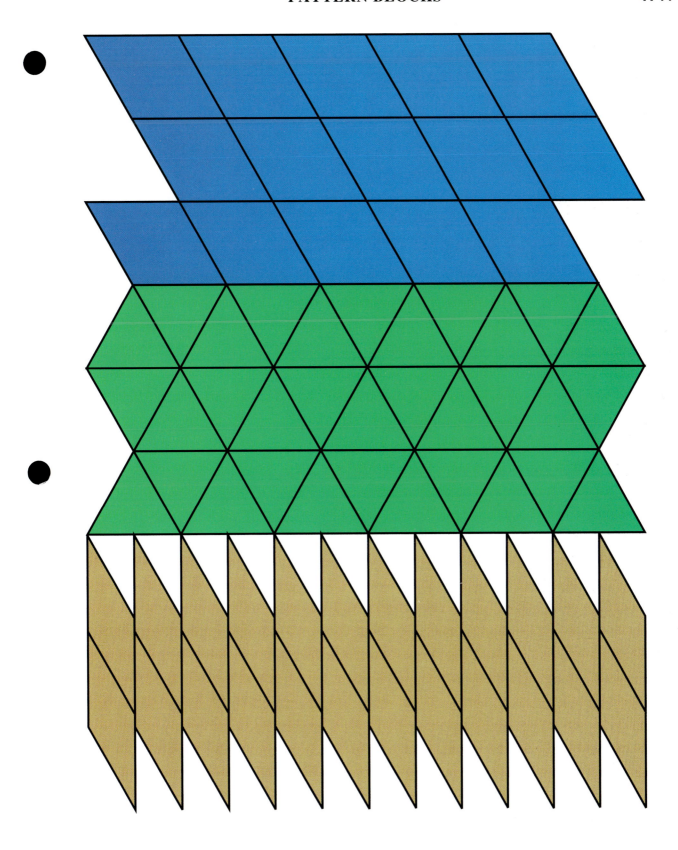

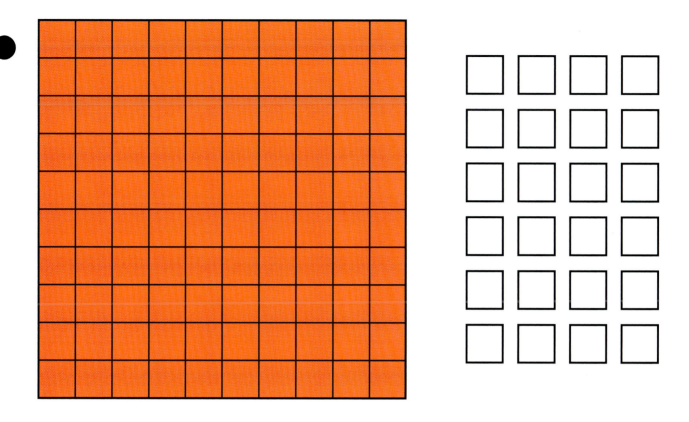

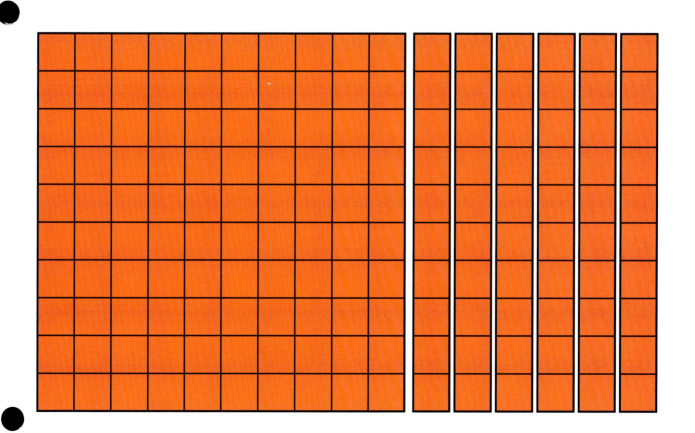

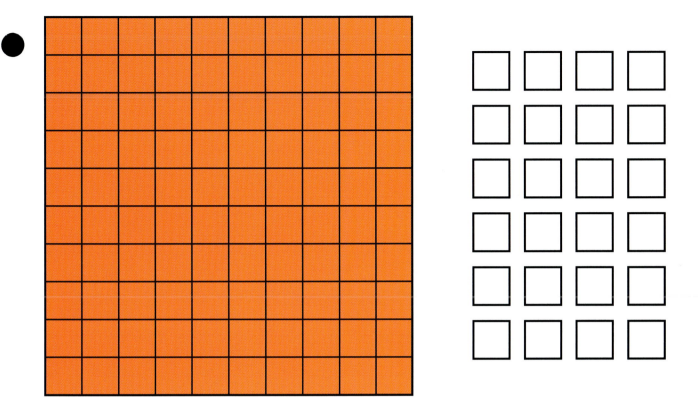

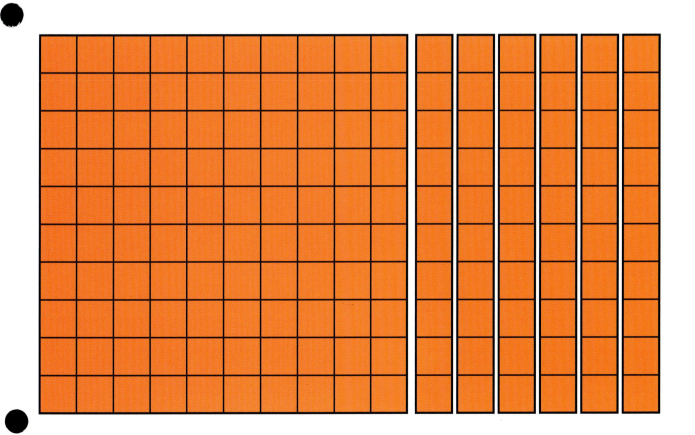

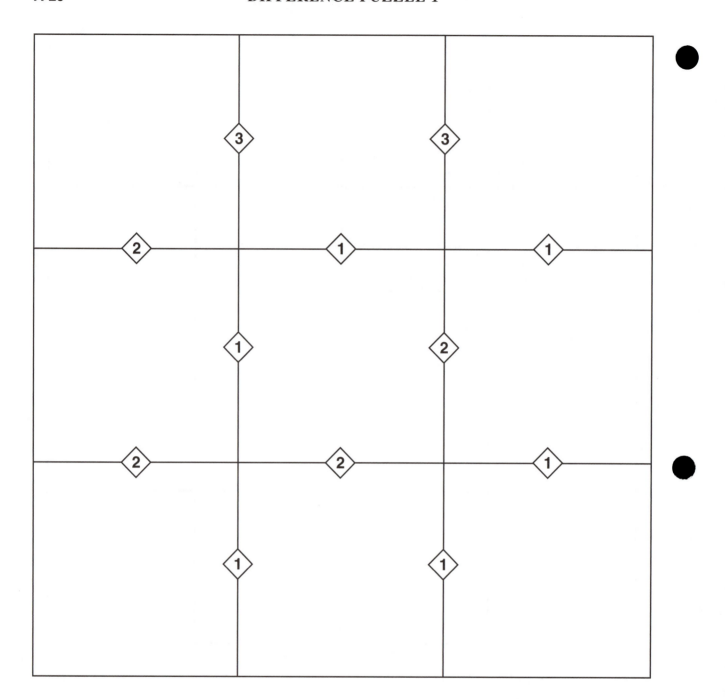

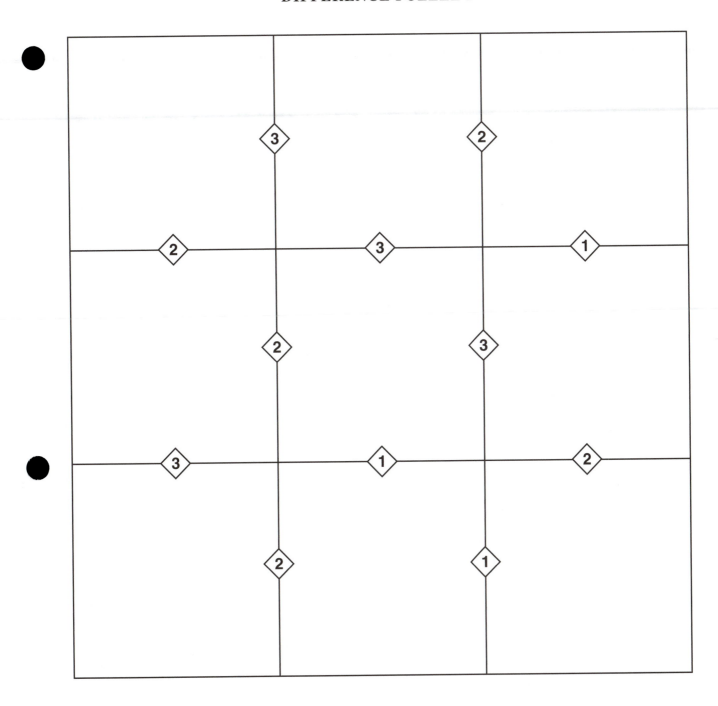

DIFFERENCE PUZZLE MASTER

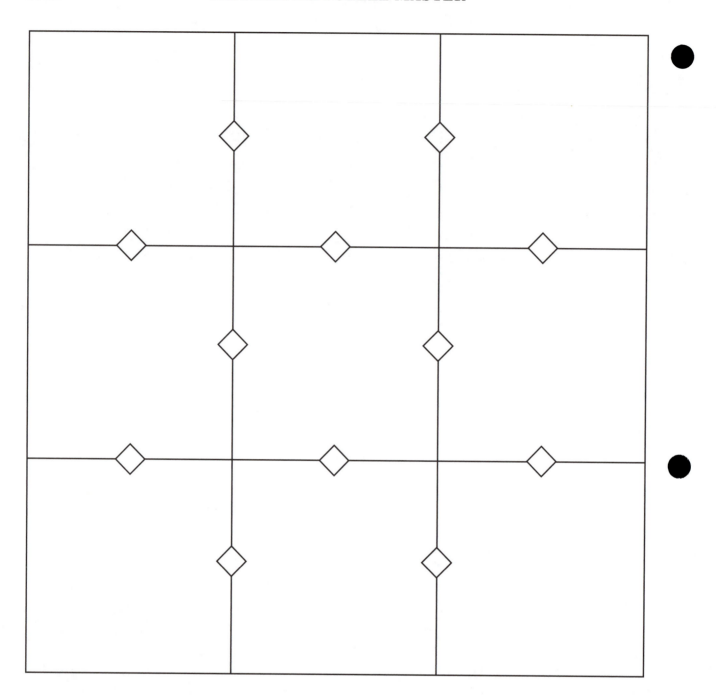

GREEN	BLUE	RED	LARGE	SMALL
NOT BLUE	NOT RED	NOT SMALL	NOT LARGE	YELLOW
NOT	NOT	NOT	NOT YELLOW	NOT GREEN
				NOT

STRIPES	NOT LARGE	NOT GREEN	
CLOWN	GREEN	NOT RED	
HOBO	BLUE	NOT STRIPES	NOT CLOWN
SMALL	RED	NOT DOTS	NOT HOBO
LARGE	DOTS	NOT SMALL	NOT BLUE

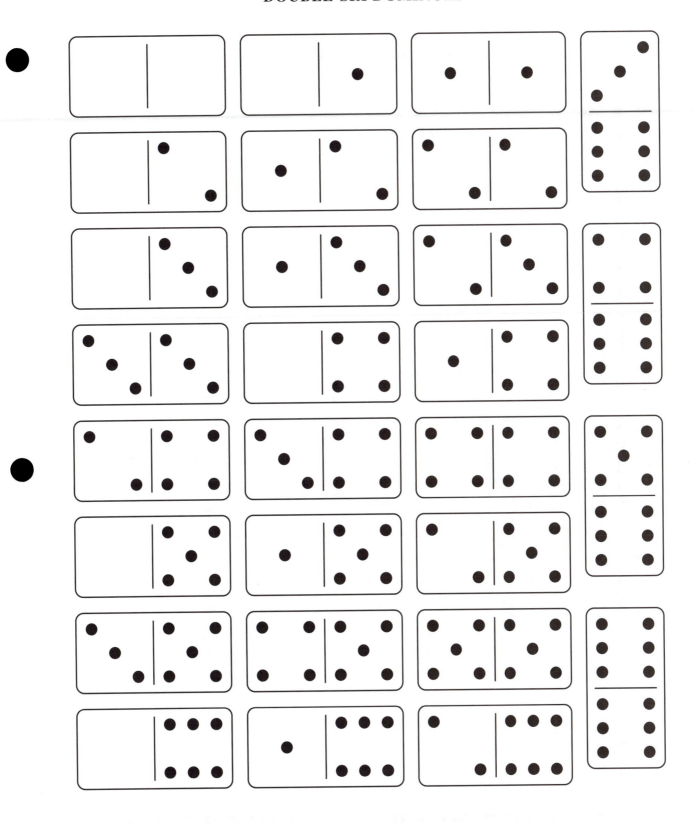

MULTIPLICATION AND DIVISION FRAME

What's My Function?

Multiply input by 3.

$y = 3x$ $f(x) = 3x$

x	0	1	2	3
y	0	3	6	9

What's My Function?

Add 4 to the input.

$y = x + 4$ $f(x) = x + 4$

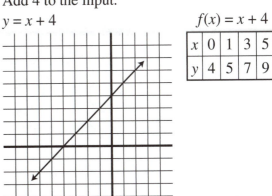

x	0	1	3	5
y	4	5	7	9

What's My Function?

Multiply input by ⁻1 and add 3.

$y = ^-x + 3$ $f(x) = ^-x + 3$

x	⁻1	0	1	3
y	4	3	2	0

What's My Function?

Multiply input by 4 and subtract 5.

$y = 4x - 5$ $f(x) = 4x - 5$

x	0	1	2	3
y	⁻5	⁻1	3	7

What's My Function?

Divide the input by 5.

$y = x \div 5$ $f(x) = x \div 5$

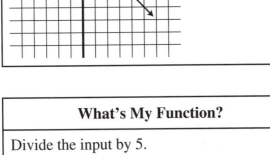

x	0	1	5	10
y	0	0.2	1	2

What's My Function?

Subtract 2 from the input.

$y = x - 2$ $f(x) = x - 2$

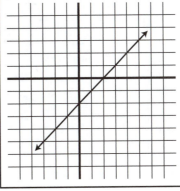

x	0	2	3	4
y	⁻2	0	1	2

What's My Function?

Divide input by 2 and subtract 1.

$y = \dfrac{x}{2} - 1$ $\qquad\qquad$ $f(x) = \dfrac{x}{2} - 1$

x	0	2	4	6
y	-1	0	1	2

What's My Function?

Add 3 to the input and divide by ⁻2.

$y = (x + 3) \div {}^{-}2$ $\qquad$ $f(x) = (x + 3) \div {}^{-}2$

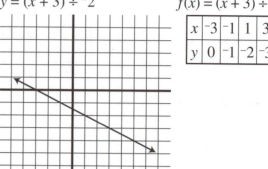

x	-3	-1	1	3
y	0	-1	-2	-3

What's My Function?

Multiply input by itself. (Square the input.)

$y = x \cdot x$ $\qquad\qquad$ $f(x) = x^2$

x	-2	-1	1	2
y	4	1	1	4

What's My Function?

Multiply input by itself and subtract 2.

$y = x \cdot x - 2$ $\qquad\qquad$ $f(x) = x^2 - 2$

x	-2	-1	1	2
y	2	-1	-1	2

What's My Function?

Multiply the input by 1 more than the input.

$y = x(x + 1)$ $\qquad\qquad$ $f(x) = x(x + 1)$

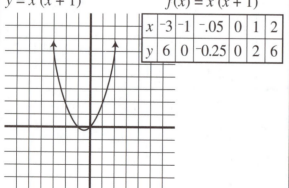

x	-3	-1	-.05	0	1	2
y	6	0	-0.25	0	2	6

What's My Function?

Multiply input by ⁻2 and add 4.

$y = {}^{-}2x + 4$ $\qquad\qquad$ $f(x) = {}^{-}2x + 4$

x	-1	0	1	2
y	6	4	2	0

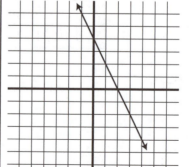

Regular dice are limited by the six faces being marked with dots to represent the numbers 1–6 only. You can construct additional cubes to be used as random number generators for developing place-value concepts as shown below.

Example:

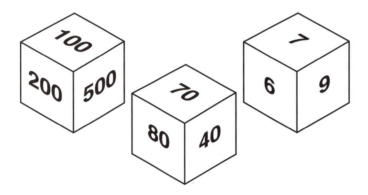

MATERIALS

Use one-inch blank wooden cubes and a permanent marker. Write the numerals to represent hundreds, tens, and units (ones) on each die.

FUTURE CLASSROOM USE WITH CHILDREN

For younger children, dice made from foam rubber upholstery padding are quiet and motivating. Mark off two-inch squares on two-inch padding. For ease in cutting, use an electric knife. When marking the numerals, go over them several times so that the ink will seep into the foam.

Fraction strips showing: 1; halves (1/2); thirds (1/3); fourths (1/4); fifths (1/5); sixths (1/6); sevenths (1/7); eighths (1/8); ninths (1/9); tenths (1/10); elevenths (1/11); twelfths (1/12).

$\dfrac{5}{6}$	$\dfrac{4}{6}$	$\dfrac{2}{6}$	$\dfrac{1}{6}$
$\dfrac{1}{8}$	$\dfrac{2}{8}$	$\dfrac{3}{8}$	$\dfrac{5}{8}$
$\dfrac{6}{8}$	$\dfrac{7}{8}$	$\dfrac{1}{10}$	$\dfrac{2}{10}$
$\dfrac{4}{10}$	$\dfrac{3}{10}$	$\dfrac{6}{10}$	$\dfrac{7}{10}$

$\frac{8}{10}$	$\frac{9}{10}$	$\frac{1}{25}$	$\frac{2}{25}$
$\frac{3}{25}$	$\frac{24}{25}$	$\frac{23}{25}$	$\frac{22}{25}$
$\frac{21}{25}$	$\frac{20}{25}$	$\frac{4}{25}$	$\frac{5}{25}$
$\frac{11}{25}$	$\frac{12}{25}$	$\frac{13}{25}$	$\frac{14}{25}$

Close to 1	Close to $\frac{1}{2}$	Close to 0

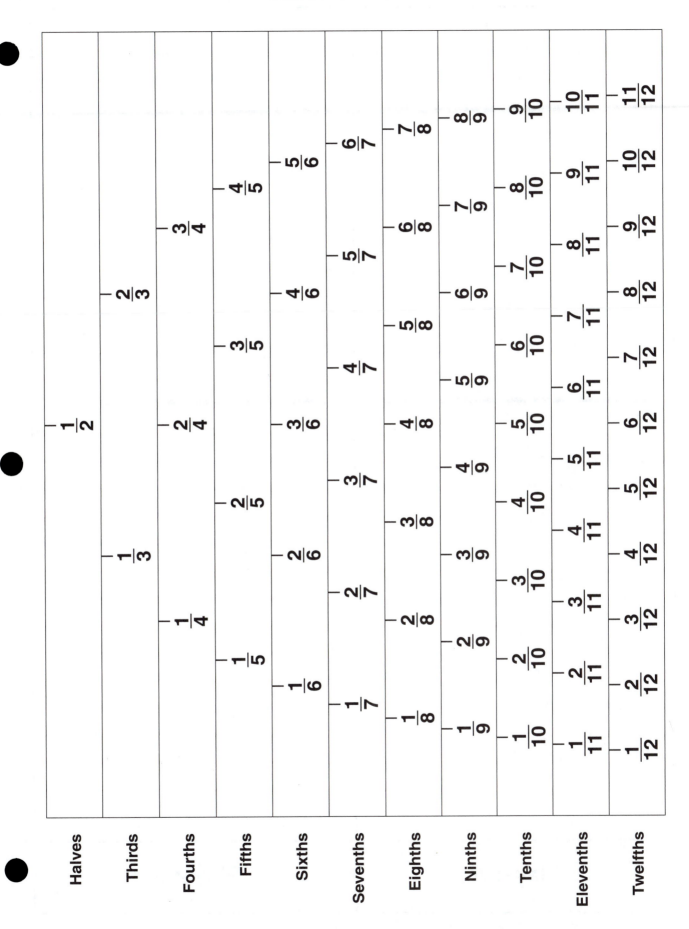

FRACTIONS GAME BOARD

_____   _____

Player A   **Player B**

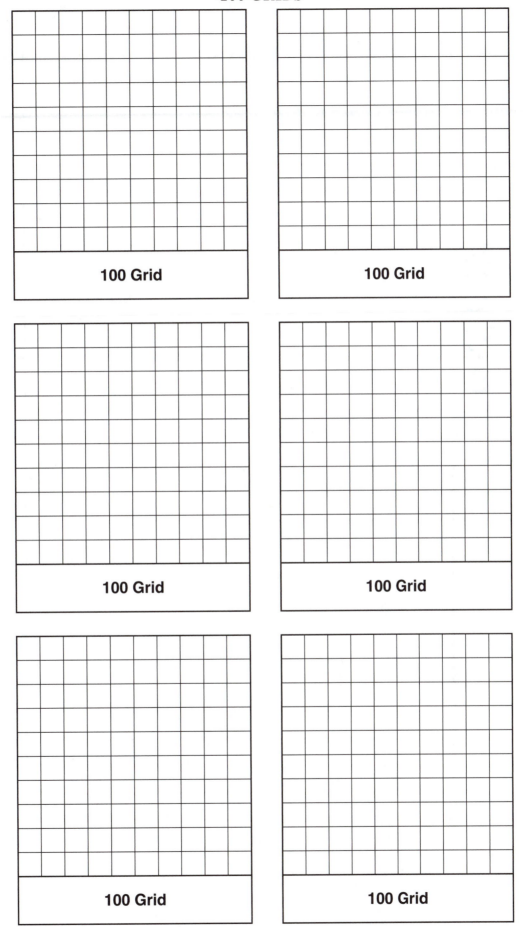

m&m's® LINE PLOT

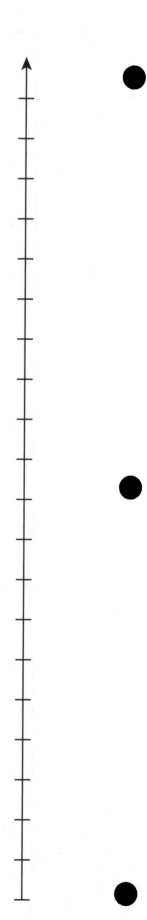

m&m's®

CIRCULAR GEOBOARDS

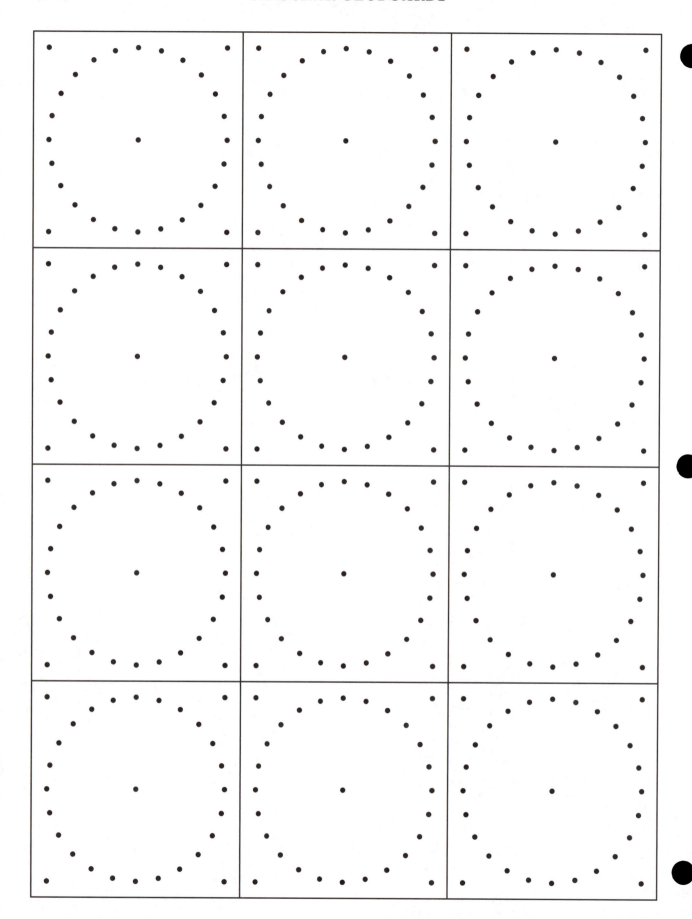

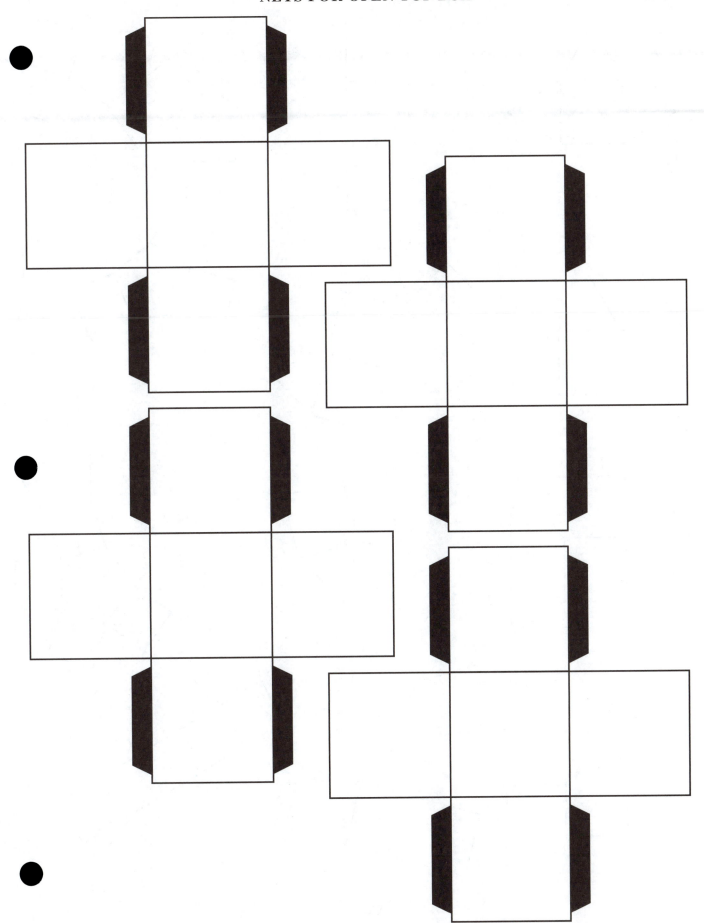

POLYHEDRAL NETS

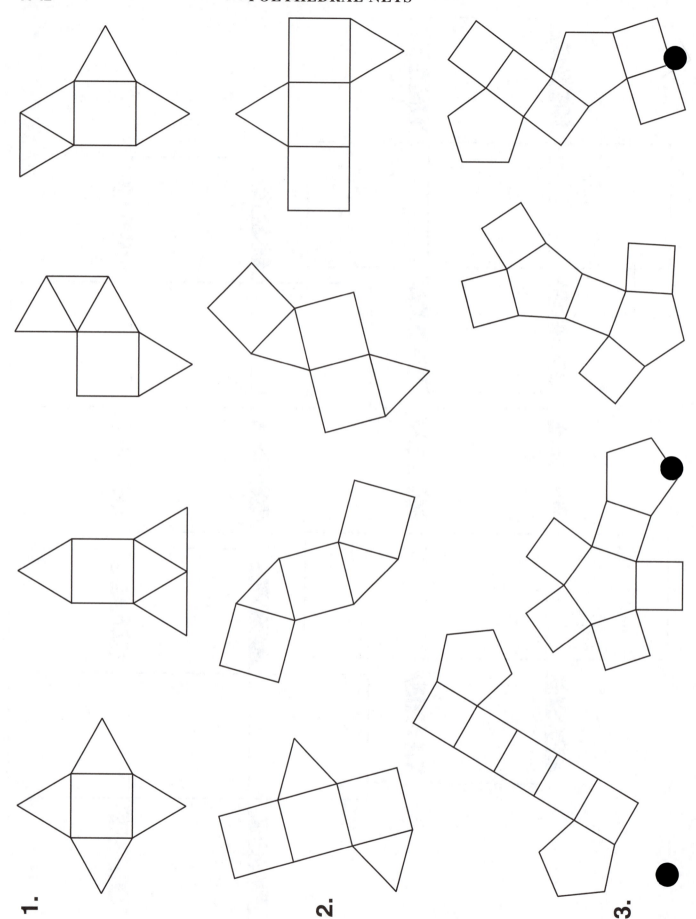

1.

2.

3.

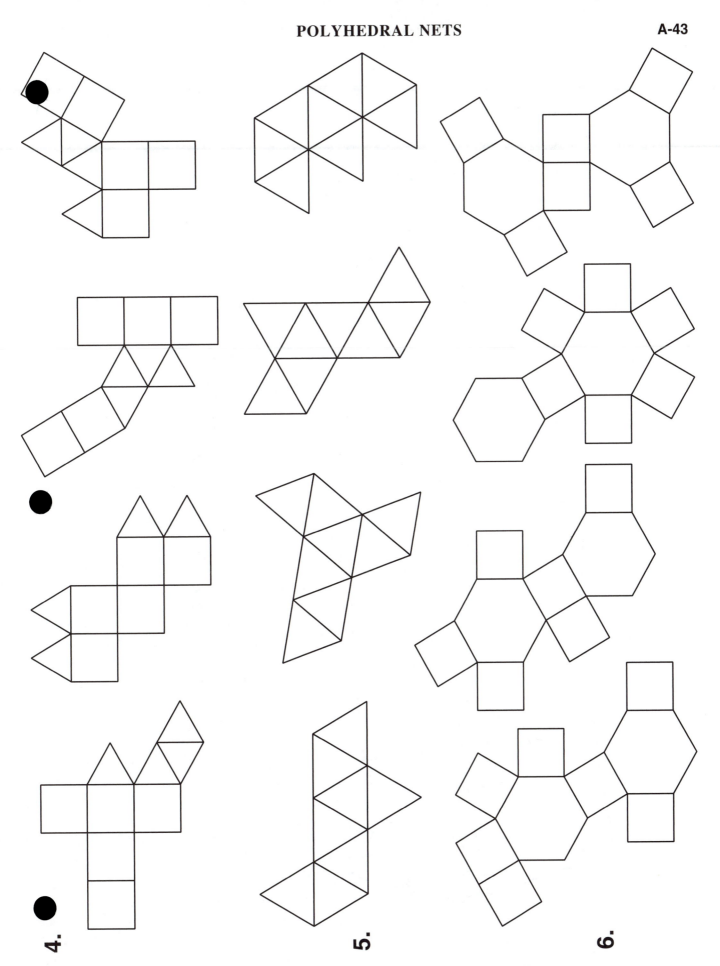

4.

5.

6.

UNITED STATES MAP

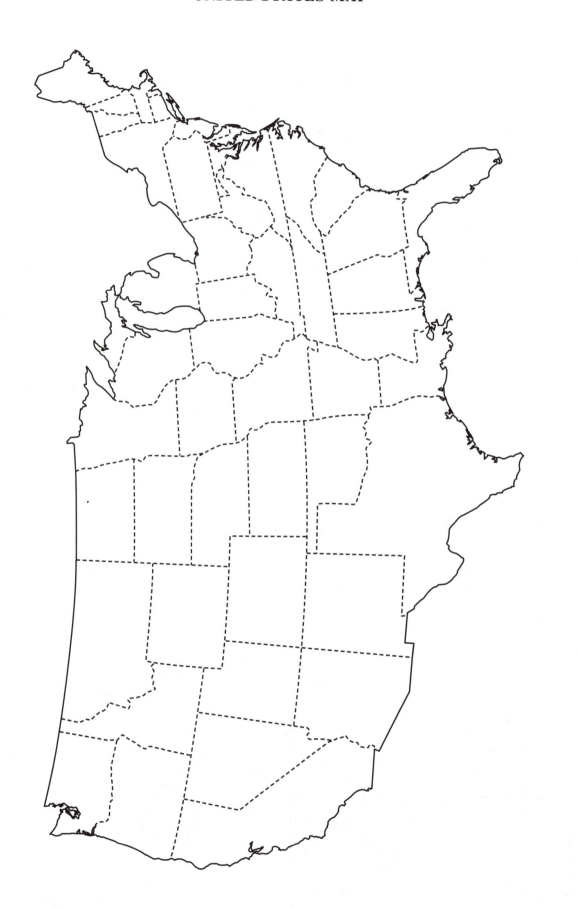

GEOBOARDS DOT PAPER

For Box 1

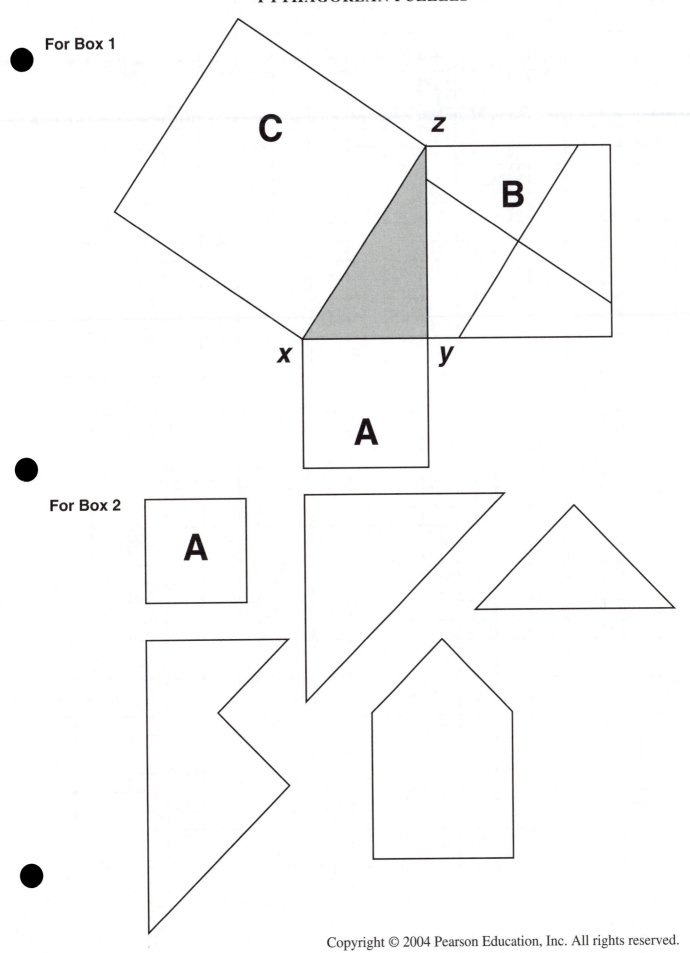

For Box 2

HALF-CENTIMETER GRAPH PAPER

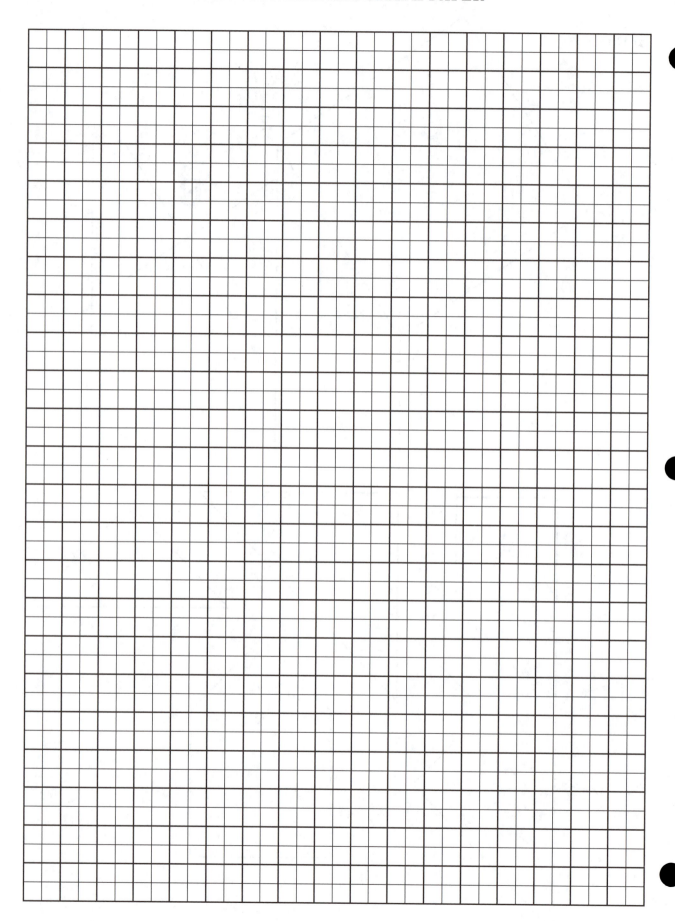

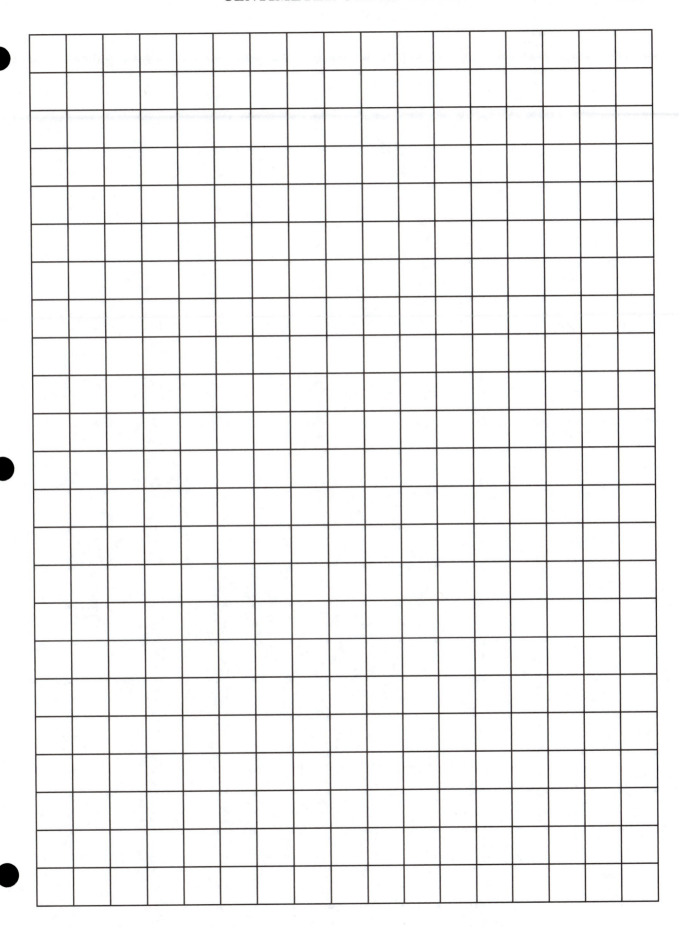

DOT PAPER

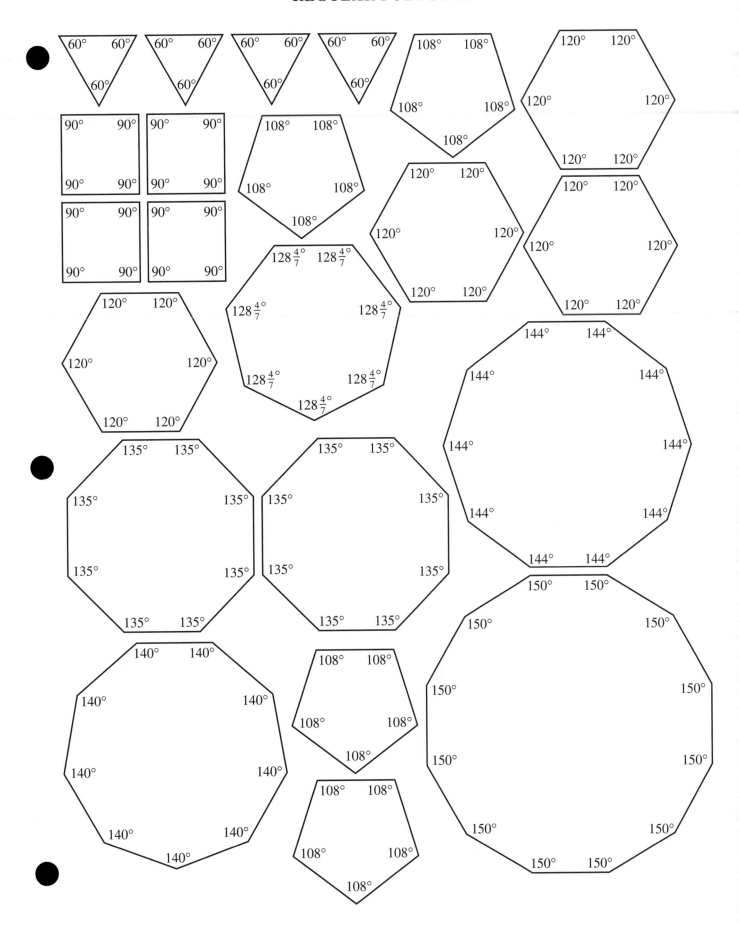